RUTH B. WINTON MEMORIAL LIBRARY
COLCHESTER HIGH SCHOOL
COLCHESTER, VERMONT 05446

WITHDRAWN

Acid Rain

Look for these and other books in the Lucent Overview Series:

Abortion	Homeless Children
Adoption	Homelessness
Advertising	Illegal Immigration
Alcoholism	Illiteracy
Animal Rights	Immigration
Artificial Organs	Juvenile Crime
The Beginning of Writing	Memory
The Brain	Mental Illness
Cancer	Militias
Censorship	Money
Child Abuse	Ocean Pollution
Children's Rights	Oil Spills
Cities	The Olympic Games
The Collapse of the Soviet Union	Organ Transplants
Cults	Ozone
Dealing with Death	The Palestinian-Israeli Accord
Death Penalty	Pesticides
Democracy	Police Brutality
Drug Abuse	Population
Drugs and Sports	Poverty
Drug Trafficking	Prisons
Eating Disorders	Rainforests
Elections	The Rebuilding of Bosnia
Endangered Species	Recycling
The End of Apartheid in South Africa	The Reunification of Germany
Energy Alternatives	Schools
Espionage	Smoking
Ethnic Violence	Space Exploration
Euthanasia	Special Effects in the Movies
Extraterrestrial Life	Sports in America
Family Violence	Suicide
Gangs	The UFO Challenge
Garbage	The United Nations
Gay Rights	The U.S. Congress
Genetic Engineering	The U.S. Presidency
The Greenhouse Effect	Vanishing Wetlands
Gun Control	Vietnam
Hate Groups	Women's Rights
Hazardous Waste	World Hunger
The Holocaust	Zoos

Acid Rain

by Rebecca K. O'Connor

LUCENT BOOKS

THOMSON
GALE

San Diego • Detroit • New York • San Francisco • Cleveland • New Haven, Conn. • Waterville, Maine • London • Munich

© 2004 by Lucent Books. Lucent Books is an imprint of The Gale Group, Inc., a division of Thomson Learning, Inc.

Lucent Books® and Thomson Learning™ are trademarks used herein under license.

For more information, contact
Lucent Books
27500 Drake Rd.
Farmington Hills, MI 48331-3535
Or you can visit our Internet site at http://www.gale.com

ALL RIGHTS RESERVED.
No part of this work covered by the copyright hereon may be reproduced or used in any form or by any means—graphic, electronic, or mechanical, including photocopying, recording, taping, Web distribution, or information storage retrieval systems—without the written permission of the publisher.

LIBRARY OF CONGRESS CATALOGING-IN-PUBLICATION DATA

O'Connor, Rebecca.
    Acid rain / by Rebecca K. O'Connor.
        p. cm. — (Lucent overview series)
Summary: Defines what acid rain is and the problem it creates for people and for the environment.
Includes bibliographical references and index.
    ISBN 1-56006-502-8 (hardcover : alk. paper)
    1. Acid rain—Environmental aspects—Juvenile literature. [1. Acid rain—Environmental aspects.] I. Title. II. Series.
    TD195.44.O26 2004
    363.738'6—dc22
                                                                2003012895

Printed in the United States of America

Contents

INTRODUCTION Not Right as Rain	6
CHAPTER ONE The Acid Rain Problem	10
CHAPTER TWO Dead and Dying Lakes	23
CHAPTER THREE Dying Forests	35
CHAPTER FOUR How Acid Rain Affects Humans	47
CHAPTER FIVE Searching for Solutions	60
NOTES	74
GLOSSARY	78
ORGANIZATIONS TO CONTACT	80
SUGGESTIONS FOR FURTHER READING	82
WORKS CONSULTED	84
INDEX	90
PICTURE CREDITS	95
ABOUT THE AUTHOR	96

Introduction
Not Right as Rain

WHEN SOMETHING IS perfect, people say that it is "right as rain." For instance, when someone is not feeling well, he or she might be told to go to bed early, to get plenty of sleep, and to expect to feel "right as rain" in the morning. It is a popular saying. People also imagine that rain is pure. Because it falls from high above, rain is assumed to be uncontaminated by the dirt and debris of the earth. In fact, when rain falls, it is thought to cleanse the landscape below.

At least that is what people have always thought. However, since the 1980s, scientists have been saying that the rain is not "right" or clean at all.

Dying forests in Europe and dying lakes across the world were commonplace stories on the news in the 1980s. At first scientists were puzzled, but research quickly revealed that the problem was with the rain. The rain was poisoned with acid, and this was causing all sorts of problems. Acid rain was poisoning lakes and trees, burning the paint off cars, and destroying monuments, and no one was sure what it might be doing to people.

Further research revealed where the acids were originating. The components of acid rain were a by-product of cars, factories, and power plants. Neither automobiles nor electricity, for example, was a luxury that society wanted to give up, but everyone agreed that something had to be done to "clean up" the acid rain.

A New Era

Acid rain was the first environmental issue to gain enough attention to urge the entire United States into action, but there were no simple solutions to acid rain. The changes that would have to be made to stop the acid poisoning would affect everyone, so politicians, scientists, and the public had to agree on a course of action.

Scientists were not entirely certain how acid rain was created, let alone exactly how the damage was occurring and how far-reaching the problem was. New theories were constantly arising. Those who studied the problem disagreed on many issues, but they all felt that more research would be a crucial part of the solution.

Politicians were having a difficult time dealing with the acid rain issue as well. Since they are voted into office, it is important that they find answers when voters make demands. And the public demands clean rain. However, the public also demands affordable electricity and transportation.

A dead fir tree in northern Russia shows the effects of acid rain, a result of pollution released from a nearby factory.

The automotive companies and the electric companies did not want new regulations that would require making expensive changes to their industries.

Thus, the biggest issue surrounding the acid rain problem was money. Cleaning up industrial emissions and vehicle exhaust would cost money. No one was exactly sure how much money would be needed, but most people felt it would be very expensive. Who would pay for this expense? No one really wanted to pay—not the government, the voters, or businesses. It was a difficult time.

Despite all of these questions and doubts, the government worked with everyone involved and passed a group of regulations called the Clean Air Act Amendments. These regulations demanded that steps be taken to reduce the pollution that was causing acid rain. The Clean Air Act also financed studies around the country, as everyone agreed that there was much yet to learn about acid rain. These regulations were the first of their kind in the United States, and people were hopeful that they would solve the problem of acid rain.

In 1999 the government revealed the results of the Clean Air Act Amendments. The results looked good. Emissions had been reduced even below the set goals, and the cost of the improvements was far less than anyone had expected. With this news, acid rain faded from media and public attention.

Damage Done

Although there are very few articles and news stories on acid rain today compared to the 1980s, it is not an issue that has disappeared entirely. The long-term research that began in the 1990s is now being published, and today people have more information about acid rain than they possessed twenty years ago. Scientists are pleased with the results of the Clean Air Act Amendments, but they also know that much more has to be done before acid rain is eliminated.

The effects of acid rain did not occur over a twenty-year period and will not go away in twenty years either. Scientists now recognize that damage from acid rain has been

Smokestacks from a power station in Australia spew pollution into the air. Acid rain poses a serious threat to forests, lakes, and ultimately people.

occurring since coal became the primary fuel for heat. In European countries, that was as early as the 1300s. Many places in the world are saturated with the acids from industrial pollution and now suffer greatly when even small amounts of acidic pollution are added to the environment.

The public may not be hearing much about acid rain anymore, but environmentalists are saying that it is as important an issue, if not more important, than it was before. Scientists know today that acid rain is slowly killing American forests and that acidified lakes will take decades to recover. There may be less pollution than in previous decades, but these scientists point out that there is still enough pollution produced worldwide to maintain the threat of acid rain on the environment. Studies are also starting to show the harmful effects of acid rain on humans.

Researchers agree that there is still much to learn about the problem of acid rain but that the world is on the right track toward a solution and recovery. However, just as the public and the government made a huge difference in the past, it will be up to them again to change policies for the future.

1

The Acid Rain Problem

THE INDUSTRIAL TOWN of Donora, Pennsylvania, had dealt with pollution for many years. The smoke and fumes from the local factories were a fact of life. The mills supplied jobs to the local townspeople, and the locals were known to state that the "smoke" put bread on their table.

In 1948 the small town of Donora became an example of the mounting problem of acid pollution. Between October 26 and 31, a dense fog settled on the town. A weather phenomenon known as inversion, when a cold air mass traps warm air near the ground, caused a disaster. Fumes from the American Steel and Wire Company's zinc and iron works mixed with the foggy air and stayed near the ground, trapped by the mass of cold air above. The poisonous air settled on the town and did not leave for nearly a week.

The townspeople of Donora realized right away that their air was tainted. The fog mixed with sulfur dioxide from the zinc and iron works and changed into a dangerous mist of sulfuric acid. A day after the fog began, patients began to crowd the two local hospitals, complaining of breathing problems, headaches, abdominal pains, and nausea.

Town officials had to shut down the city, closing offices and stopping people from driving because drivers could not see through the dark, polluted fog. Ambulances were even stopped from driving through the town. Firefighters traveled on foot, patrolling the streets and trying to help citizens, some firefighters even dragging 130-pound oxygen tanks door to door. At nearly every house they found people wheezing a little or unable to breath altogether. They gave each person a couple of breaths of clean air from the tanks and then moved on to the next house, helping as many residents as they could.

By the third day, people were dying. Doctors recommended that citizens with breathing problems leave the town, but driving was prohibited due to the lack of visibility. Evacuation was impossible. Nearly six thousand people suffered the effects of this fog, and twenty people, five women and fifteen men, died. These deaths were due to respiratory failure, meaning the victims were unable to breathe. Many others would suffer the effects of breathing in the poisoned air throughout their lives. These people would suffer from asthma and have difficulty breathing for as long as they lived.

The people of Donora that survived the fog were worried and angry. They knew that their air was polluted, but they had not realized that it could be deadly. Local residents wrote to

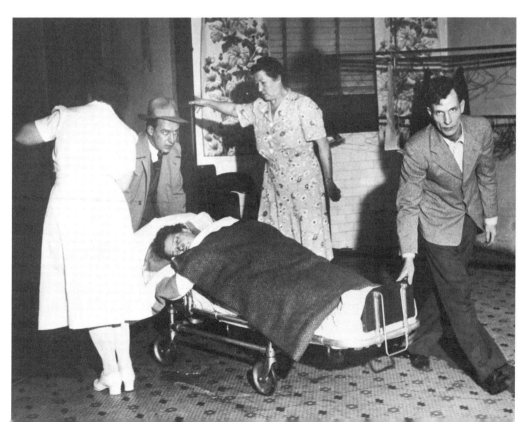

Medical personnel rush a victim of the deadly Donora fog to the hospital. In October 1948, a fog mixed with acid pollution from a local zinc mill blanketed Donora and killed twenty people.

the governor, demanding that action be taken to keep a similar incident from happening again. One such letter, from Lois Bainbridge, can be found in the Pennsylvania state archives. Bainbridge felt that the zinc works were to blame for the poisoned air and suggested that the governor shut them down. She believed this would "do away with that awful acid and smoke. It eats the paint off your houses. Even the fish cannot live in the Monongahela River, the bank on which the mill is situated."[1]

What Bainbridge was witnessing was acid pollution from the zinc mill. When the mill burned metal compounds called ores to separate out the usable zinc, the by-product was pollution. Smokestacks from the mill pumped streams of smoke full of poisonous sulfur dioxide into the air. Despite Bainbridge's request, the mill was not shut down until 1956, and it would be over twenty years before the U.S. Congress had hearings regarding the problem of pollution. In 1970, when federal legislators took action, the town of Donora would be used in congressional hearings as an example of the problem of acid pollution.

Acid Pollution

As the smoke from the Donora zinc mill illustrates, acid pollution is the introduction of harmful acids into the environment by human activities. Pollution is the addition of anything that is harmful to the environment. Acids, although they naturally occur in the environment in small amounts, become pollution when they are introduced in large amounts.

Acid pollution is created mainly by the burning of fossil fuels. Fossil fuels are resources retrieved from within the earth and under the seabed. These fuels are common in everyday life and include oil, gas, and coal. These fuels are also indispensable to modern society. As writer John McCormick notes, "the world now relies on fossil fuels for 90 percent of its commercial energy supplies."[2] Fossil fuels are burned to release the energy that is stored within them. When burned, these fuels release not only energy but gases as well.

There are several common ways that fossil fuels are used and contribute to pollution. The burning of gasoline in vehicles creates emissions that add acids to the environment. Smelting, or roasting, sulfur compounds to separate out pure metals such as zinc, nickel, and copper produces pollutants as well. Burning coal, natural gas, or oil to heat homes also adds to the pollution problem.

By far the most common cause of acid pollution is the burning of fossil fuels to produce electricity. Electricity is produced in U.S. power plants when fossil fuels are burned to heat water and make steam. The highly pressured steam is aimed at the blades of turbines to make them spin. The spinning creates electricity. The demand for large amounts of electricity means great quantities of fossil fuels are burned, and that, in turn, produces much pollution.

The pollutants created from the burning of fossil fuels are nitrogen oxides and sulfur dioxide. Coal, especially coal that has a high sulfur content, gives off sulfur dioxide. The burning of oil and gas produces nitrogen oxides. Both these pollutants are poisonous gases that, when expelled into the air, mix with the moisture in clouds and change into acids.

Scientists are not certain exactly how nitrogen oxides and sulfur oxides turn into acids, but they know that this change only happens when they become airborne. Scientists also know that sunlight, moisture, air currents, and oxygen all play a role in the change. Depending on the presence or absence of these elements and the amount of time the pollutants are in the air, eventually the pollution becomes nitric and sulfuric acids.

Once sulfur dioxide and nitrogen oxides change into acids, they must at some point return to the earth. Scientists call these falling acid particles "acid deposition." Deposition is when a material is deposited on the ground, buildings, water, or even people. There are two kinds of acid deposition, dry deposition and wet deposition. Dry acid deposition occurs when acids return to Earth as gas or small dry particles. Wet acid deposition occurs when acids return to Earth as snow, hail, fog, and the most widely known form of acid deposition, rain.

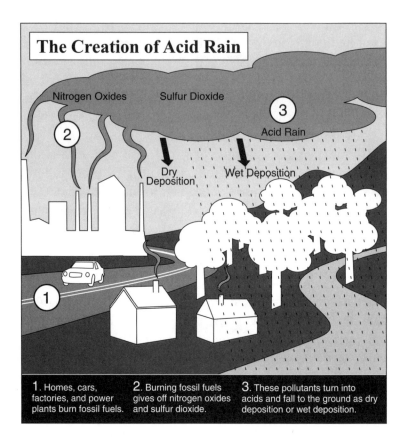

1. Homes, cars, factories, and power plants burn fossil fuels.
2. Burning fossil fuels gives off nitrogen oxides and sulfur dioxide.
3. These pollutants turn into acids and fall to the ground as dry deposition or wet deposition.

Wet acid deposition is the most destructive form of deposition. A moderate rain will remove more acid from the air in one hour than could fall in dry form in over two or three days. This high concentration of acids falling in the form of rain has become an environmental issue in many countries and is commonly called "acid rain" even if it does not always fall as rain.

Acid Rain

Although it may seem that acid rain is a product of recent times, the phenomenon has been around for more than a hundred years. A Scottish chemist named Robert Angus Smith, studying the rainfall in and around his hometown of Manchester, England, found the rain to be unusually acidic in 1852. The phrase "acid rain" was coined by Smith in a book he published on these findings in 1872. According to

John McCormick, Smith "warned that plants and materials were being damaged"[3] by acid rain. At the time, his views that pollution from factories was causing dangerous acid rain were not accepted by the scientific community, but today scientists agree that industry is a part of the problem, that rain with abnormally high acidity can cause environmental damage.

Unpolluted rain is normally somewhat acidic. This is because acids are a natural part of the environment. Acids can be added to the environment by volcanoes or rotting vegetation. Even carbon dioxide, the gas that animals—including humans—exhale when they breathe, contributes to the normal acidity of rain. The acid in rain does not become a problem until the level of acid or its acidity rises and becomes excessive.

Acidity in the simplest terms is sourness. The sourer a solution is, the higher its acidity is. This can be explained by looking at simple household items. For example, milk is only slightly acidic, and tomato juice slightly more acidic. Lemon juice is very acidic, and battery acid is extremely acidic. Battery acid is so acidic that it can burn skin. However, scientists need to be as precise as possible when measuring acidity in waters and soils, so they measure sourness on the pH scale.

The pH values can range from 0 to 14. The numbers below 7.0 indicate acid content. The lower the number is, the higher its acidity. For example, lemon juice has a pH of 2.0, and battery acid has a pH of 1.8. Unpolluted rain has a pH of 5.6, and anything with a lower pH is considered acid rain.

A rain shower with a pH of 5.6 is not harmful, but in many parts of the world, rain with a much lower pH has fallen. In parts of Pennsylvania, West Virginia, and New York, showers with a pH reading between 3.0 and 4.0 have been recorded. The lowest readings recorded occurred in 1978 in Wheeling, West Virginia, where the pH of the rain was measured at just under 2.0. This level of acidity is comparable to a shower of lemon juice. When highly acidic rains fall, they can be damaging.

Acid Rain Damage to Materials

Acid rain has far-reaching consequences and affects almost everything it touches. Although acid rain has never been recorded at a high enough acidity to burn human skin, years of acid rainfall can etch stone and destroy ecosystems. As rain becomes more acidic, plants, animals, and even buildings and monuments are affected.

By studying tombstones, researchers have discovered the intensity of acid damage to stone. Gravestones that have been exposed to high levels of acidity wear away, leaving the words carved in the stone unreadable. Tom Meierding, a researcher from the University of Delaware, discovered a cemetery in Leadville, Colorado, with gravestones that were in very bad shape. The words on these stones were barely readable, and the stone was pocked and worn away. He stated that he was not surprised to discover "that at one point in time, Leadville had been the smelter capital of the West, processing gold, silver and other ores mined during the Gold Rush era."[4] The acid pollution from the smelters had mixed with the air and burned away the words on the tombstones.

Stone is not the only material that can be affected by acids; metals can be damaged as well. Many treasured metal statues around the world are slowly being destroyed by acid pollution. One of the most famous American landmarks, the Statue of Liberty, has received a lot of acid damage over the last few decades. In the late 1980s, the Statue of Liberty underwent extensive repair. It was noticed that there were many black spots and thin spots on the statue. Experts concluded that this damage was due in part to acid rain.

Important monuments all over the world are being impacted by acid rain. The Taj Mahal in India, the Parthenon in Greece, and even the brilliant stained glass windows of European cathedrals are slowly being worn away by acid rain. Scientists say that ancient buildings and sculptures in many cities have weathered more during the last twenty years than in the last two thousand years.

Acid can do more than just burn the surface of materials;

it can also corrode and destroy them. Metals such as steel, copper, nickel, zinc, lead, and cast iron can all be corroded to varying degrees; some more quickly than others. According to John McCormick, "Studies in the Northeast United States have found that rates of corrosion to galvanized steel are almost three times greater in urban areas than in the Adirondack Mountains"[5] where there is much less pollution. Some scientists feel the most disturbing example of this corrosion from acid rain is the collapse of the Ohio River Silver Bridge.

The Silver Bridge was built in 1928, connecting Point Pleasant, West Virginia, with Kanauga, Ohio. The bridge was called "Silver Bridge" because of its shiny aluminum paint. It was a commonly used bridge and often very busy during rush hour.

Acid rain is so powerful that years of exposure to it can destroy stone sculptures like this one in England.

On December 15, 1967, during Christmas shopping rush hour, disaster struck. There were so many people returning from shopping and other Christmas errands that traffic was at a complete standstill. Cars were lined up on the bridge waiting to move forward. With no warning the bridge buckled and collapsed, sending thirty-one vehicles and sixty-four people plummeting into the Ohio River.

The bridge's collapse took less than one minute. Chris LeRose of the West Virginia Historical Society states that witnesses who saw the bridge collapse said, "it looked like the bridge fell like a card deck."[6] The people on the bridge had no time to get out of their cars and off the bridge. Many witnesses tried to help victims who fell into the water, but there was little they could do. Rescue crews arrived and were able to save the people who had escaped their vehicles from drowning in the Ohio River. Vehicles fell into the water and sank so quickly that many of the victims were unable to get out of their sinking cars and trucks. Eighteen of the people that were on the bridge survived, but forty-six died.

Puzzled about the cause of the collapse, engineers spent two years reconstructing the bridge out of the wreckage to find out what had happened. Finding every tiny piece of the fallen bridge, they took the wreckage to a reconstruction site and carefully put it back into its original form.

When the bridge was pieced together, the discovery was made that a single joint had corroded and broken under the stress of the traffic. The failure of this single joint began a chain reaction as one joint after another broke, quickly destroying the bridge. This first joint had a small flaw in it, and years of stress from normal use in addition to the corrosive acid pollution from factories on the Ohio River caused its eventual failure. Today bridges are carefully inspected for the acid corrosion caused by acid rain.

Acid Rain Damage to Living Things

Acid rain can also change the balance of natural acids in the environment, and this can be harmful to life that requires

acid levels to remain constant. As acidity rises, animals and plants die. This is perhaps most apparent in aquatic environments where the acidity of the water can change dramatically when acidic precipitation is added. According to researcher Kim J. DeRidder in the book *Acid Rain and Friendly Neighbors*, "Damage from acidification of [lakes and rivers] is often defined as degeneration, reduction or extinction of species populations of fish, reptiles, crustaceans, microbiotic life, insects, or aquatic vegetation."[7]

One example of acid rain's effects on such environments can be seen in Virginia's Paine Run River, located in Shenandoah National Park. In April 2001 this stream was listed as one of the nation's most endangered rivers. Ellen Baum of the Clean Air Task Force states this is due to the fact that "without further cuts in nitrogen and sulfur, Paine Run will become too acidic to sustain populations of brook trout and other aquatic organisms."[8] Brook trout cannot survive once the pH reaches 5.0. A normal stream has a pH of around 6.0, but the Paine Run River's pH has dropped near

In December 1967 the Silver Bridge on the Ohio River broke in half when a joint corroded by acid rain gave way. The disaster claimed forty-six lives.

5.0. Scientists are watching the river's pH level closely because when it reaches 5.0, the fish in the river will begin to die. A normal stream with a pH of 6.0 may have nine thousand fish per stream mile, but where acid rain has reduced the pH below 5.0, there may be no fish at all.

The Paine Run River is one example of many places that are troubled by acids in the environment. There are similar examples all over the world. Until recently acid damage was confined to industrial cities. Today, acid rain even damages places far away from where pollution is created.

Sharing the Pollution

Sulfur and nitrogen pollution has caused problems in industrial areas for over a hundred years. The killer fog in Donora, Pennsylvania, is merely one example. According to John McCormick, "The litany of the most infamous 'killer smogs' has now become all too familiar: London 1873 and 1880 (2500 deaths); Meuse Valley, Belgium, 1930 (63 deaths); . . . London, 1962 (340 deaths)."[9] Many people who survive these poisoned fogs still suffer the effects. Diseases such as asthma, bronchitis, and pneumonia are far more common in these cities than in places that do not have industries burning fossil fuels.

In 1970, in an effort to combat the problems of this pollution, Congress passed a bill called the Clean Air Act. This act was a group of laws that set strict guidelines for the amount of pollution that industries could release into the environment. Congress demanded that measures be taken to ease the pollution surrounding factories and power plants. The death of people in industrial cities around the world proved that people must have cleaner air.

In order to meet requirements reducing local air pollution, companies built tall smokestacks. Tall stacks release the pollution produced by the factory much higher into the air; pollution is then spread over a greater area and is not as concentrated. The tallest and most successful of the stacks were over 900 feet (274 meters) high and were called "superstacks." The most famous of these stacks was one built by a company called International Nickel Company (INCO) in Canada. This superstack was 1,552 feet (473 meters) high and contributed dramatically to the improvement of air quality in and around the city of Sudbury, where the factory was located. At first it seemed like a great success. The air in towns and cities near this and other superstacks became much cleaner, but there were consequences to sending the pollution high into the atmosphere.

The most troublesome consequence of superstacks was that now the pollution that was being pumped into the air could travel on the winds. Cities were sharing pollution with the countryside and forests. Researcher Brian A. Forster noted, "It is unfortunate that government policies designed to solve one environmental problem may have generated another problem."[10]

A Worldwide Problem

Pollutants lifted into the air by tall smokestacks could be carried on prevailing winds. These winds can carry pollutants for hundreds of miles, depositing acids far from their

A superstack at Canada's International Nickel Company pumps smoke high into the air. The pollution from this factory affects forests and towns hundreds of miles away.

original source. Studies show that acid rain episodes in North Dakota and South Dakota, for example, are a result of pollution from copper smelters in Arizona and New Mexico. Studies also show that 80 percent of Colorado's acid rain has come from outside the state boundaries.

This sharing of pollution across local and national boundaries is happening all over the world. Much of the pollution in Canada is generated in the United States. Recently many Asian countries have begun to suffer the effects of pollution from China. However, the most troubling case is in Sweden.

In the 1960s and 1970s, Sweden began having a particularly difficult problem with acid rain. Although the country did not have the kind of coal-burning factories that create sulfur pollution, there was a lot of acid rain falling on the country. Dying fish were reported in lakes all over western Sweden, and the cause was believed to be acid rain. Sweden reported that acid rain was also damaging soils, constructions, and monuments and placing human health at risk.

A concerned scientist named Svante Oden began researching where this pollution was coming from. In 1967 through 1968, Oden presented evidence that the rising acidification of Swedish lakes was due to air pollution originating in Great Britain and western and central Europe. This trend continued, and even today Sweden receives 58 percent of its sulfur pollution from other countries.

Although Sweden produces very little pollution, the country has been drastically affected by traveling pollution from other nations. About ten thousand of Sweden's eighty-five thousand lakes are now so acidified from the effects of acid rain that some species cannot survive in them. Today acid rain is a worldwide problem, not only in industrial countries but even in countries that create very little pollution, like Sweden.

2

Dead and Dying Lakes

ONE OF THE most obvious effects of acid rain was dead and dying lakes and waterways. A dead lake is created when the acid levels change the pH of the water and kill off the creatures that inhabit the lake. Lakes and streams suffering from these effects were being discovered all over the world in the 1980s when the issue of acid rain came to the forefront of environmental concerns.

A Tale of Two Lakes

In an effort to discover exactly how acid rain destroys lakes, researchers at the University of Wisconsin-Madison began a landmark study on the effects of acid rain. The subject of their research was a lake called Little Rock in Wisconsin. Scientists proposed introducing acid into the lake to study its effects.

The idea of intentionally destroying a lake was met with a great deal of resistance. Scientists argued that in order to learn how to stop acid rain, they had to understand exactly how it damaged water systems and how long it would take for the waters to recover. Little Rock Lake would be perfect for the study because of its unique shape, which would allow scientists to divide it in half. Finally the state legislature agreed to let the study begin.

In 1984 scientists separated the hourglass-shaped lake into two sections with a mesh curtain, keeping one side in its natural state while the other side was slowly acidified.

The original pH of the lake was 6.1. Scientists lowered the pH every two years until the lake had a pH of 4.7. Then scientists allowed the lake to recover on its own.

Thomas Frost, the director of the study, stated, "It essentially became a tale of two lakes, as the character of the acidified water began to dramatically change." The difference between the two lakes became dramatic, and Frost noted, "We found that the pH levels had a controlling but indirect influence for nearly every biological factor in the lake. The nature of the food web completely changed."[11]

Scientists were able to watch the portion of the lake that had been altered with acid slowly die. Sport fish (fish that sportsmen catch, such as bass and perch) survived the change, but most other life did not. The offspring of the bass and perch were among the first victims. The zooplankton in the lake, a tiny organism used by many animals for food, was greatly affected. Some species of the plankton died off almost completely, while others flourished. As other small organisms, like bacteria, died, the lake water became crystal clear. Although the crystal clear water looked beautiful, it was a sign that life was leaving the lake.

Then thick algae, called sphagnum moss, took over the bottom of the lake. Researchers nicknamed the algae "elephant snot" because it was so thick and slimy. The algae choked any natural plants that tried to grow at the bottom of the lake, killing them. In comparison, the lake on the other side of the barrier thrived and remained the same as it had always been.

In 1990 scientists stopped putting acid into Little Rock Lake, but they continued to study the lake, waiting to see if the lake could recover on its own. By the year 2000 the pH level of the lake was back to normal, but the plants and animals in the lake were not. Most of the fish and plants that had previously lived in the lake existed in fewer numbers or had not returned at all.

Scientists were able to prove that acid damages lakes and that recovery takes a long time. The Little Rock Lake study is considered a landmark study because scientists can use the information gained from their research to understand what is happening to lakes around the world. However, not all lakes are affected exactly as Little Rock was.

Lakes at Risk

Not all lakes are at risk from acid rain. Even in places where the rain is highly acidic, some lakes are able to absorb the acid without being harmed. This ability to buffer acids is called a lake's acid neutralizing capacity, or ANC. When scientists study a lake, they look at the lake's water source, composition of rock on the lake floor, and lake size to decide what the ANC is and the degree to which the lake is at risk from acids.

In nature there are naturally occurring substances that buffer, or weaken, acids. Limestone is one such material. Lakes that have rock bottoms, or bedrocks, made from limestone have a high ANC and are able to neutralize acids that are added to the waters. Other lakes that have granite bedrock, for example, which do not release any neutralizing chemicals, are unable to counter acid rain effects and quickly become acidic. Such lakes have a low ANC.

Lakes that receive water from streams and groundwater are also less likely to become acidic and have a high ANC.

Pristine lakes can exist even where rain is highly acidic. Lakes with limestone bottoms, for example, have the ability to absorb acid without being harmed.

Streams and groundwater often carry buffering substances from the soil. As fresh water flows into the lake, the lake is able to buffer acid rain. Lakes that receive most of their water from precipitation falling directly on their surface are most at risk of acidification.

The size of a lake will also determine how quickly it becomes acidic. It takes longer for a larger or deeper lake to become acidic. The pH of small and shallow lakes will rise more quickly.

Although many lakes are naturally resistant to acids, there are many others that are not. In the northeast United States and in Canada there are thousands of lakes that rest on granite. Many of these lakes are small and only receive water from rain and melted snow. According to Ellen Baum of the Clean Air Task Force, "Forty one percent of lakes in the Adirondack region of New York are either chronically or episodically acidic. The same holds for fifteen percent of the lakes in New England." Ellen Baum also states, "Acid rain has resulted in large losses of fish and aquatic communities in over 30,000 sensitive lakes in Ontario and Quebec."[12]

Many European lakes suffer from the same problems as those in North America. Those effects are most notable in

A lake's size determines how quickly it becomes acidic. Small bodies of water like this one become acidic more quickly than large or deep lakes.

Scandinavia. According to the Swedish Environmental Protection Agency, "In Scandinavia most soils are poor in limestone and are therefore more vulnerable to acidification than those in most European countries."[13] Currently there are ten thousand lakes in Sweden that are so acidified that sensitive organisms cannot live in them.

There are lakes all over the world that share similar characteristics, like granite bedrock, which cause them to be sensitive to acids. Once scientists realized that a lake's characteristics are a critical part of acidification, they began to research how lakes become acidified. Researchers discovered that the story of acidification is similar around the world.

Acidification

In lakes and streams that have a low acid neutralizing capacity, acidification can happen all at once or slowly over time. Regardless of the rate of acidification, the results are the same. Life in the acidified water dies.

Once a lake or stream's pH reaches 5.0, it is considered acidic, and this acidity changes the natural state of the water's ecosystem—the way in which the plants, animals, and their environment relate to ensure the survival of life in that specific locale. Life in a lake, for example, is dependant on the lake's ecosystem remaining the same. Even small changes can damage all the life that lives in the ecosystem. In order to understand how this happens, it is important to understand how an ecosystem works. An ecosystem is the balance of life within an area. This area can be small, like a pond, or large, like a lake.

All living things in an ecosystem depend on one another for survival. In a lake this includes tiny organisms like zooplankton and bacteria as well as insects, fish, frogs, and birds. Each of these animals is an important part of the food chain. For example, fish eat the insects, birds eat the fish, and if one of these animals dies, bacteria "eats," or breaks down, its body. If any of these elements of the food chain is missing, the whole ecosystem may suffer. If the insects disappear, the fish will not have a food source; if the fish die, the birds will not have food, and so on.

When one species of animal is destroyed by the rising of acidity of a lake, the whole ecosystem will be disrupted and may even be destroyed. As the pH rises, more links in the food chain disappear and the ecosystem falls apart even further. Eventually there is nothing in the lake but crystal clear water.

How Acid Affects Aquatic Life

Animals and plants that live in water are considered aquatic, and acid rain impacts all aquatic life differently. Some species deal well or even flourish when acid is introduced into an aquatic environment. Other species are very sensitive and die. Regardless, either type of change is harmful to an environment.

The first organisms that die from rising pH levels are plankton and invertebrates, which are animals without a backbone, like insects. Animals that are especially small, like the plankton, are much less tolerant of changes to their environment. Insect larvae, clams, and young crayfish are also quickly affected by acidification. These animals have soft bodies, and the acid kills them by burning their thin skin.

Amphibians are also susceptible to harm from acidification. Many amphibians breed in small ponds that are created from snowmelt. These ponds can be highly acidic due to their size and the water source. This acidity has been known to cut down on frog populations that live in these small ponds.

Although the acid does not affect many adult frogs, it is very dangerous to frog eggs. Frog eggs are not like bird eggs or reptile eggs, because they do not have a shell. Without a shell, acids in the water can easily enter the frog egg. The acids then damage the embryo, leaving the young frog unable to develop. According to Professor Bruce Forster, "evidence suggests that many species [of frogs] suffer 50% mortality at . . . pH values 5.0–7.0."[14]

Some aquatic creatures are more susceptible to acidification than others. High acid levels rarely affect adult frogs like this one but are very dangerous to frog eggs.

Fish are perhaps the most obviously affected animals in an aquatic ecosystem. The disappearance of fish is the first thing that humans notice in a dying lake. Unfortunately, by the time all the fish are gone, the lake is usually completely dead.

The Lumsden Lake Experiment

One of the most famous cases of dying fish in an acid lake was recorded by Harold Harvey, a biologist from the University of Toronto. In the 1960s, Harvey released four thousand tiny salmon, called fingerlings, into Lumsden Lake, in Ontario, Canada. The lake was sixty feet deep and cold, just the right conditions for the salmon to survive. Harvey hoped that the salmon would thrive and breed in the lake so that they would be available to fisherman.

The lake that Harvey chose seemed like a perfect place for the salmon. Not only was it the perfect depth and temperature, but it was also located deep in the wilderness. There were no nearby towns, roads, or highways to pollute the lake. It was remote enough to be clean and unspoiled.

According to government fish and game statistics, Lumsden Lake had eight species of fish in it already, proving that it was a good habitat for fish.

When Harvey returned to the lake in 1971 to check on his salmon, what he found puzzled him. Harvey and a team of researchers used large nets to scoop fish out of the lake, but they could not find any salmon. They did not catch any of several other species, such as trout, herring, or perch, in the lake. Instead, the only fish they found were white suckers.

There was also something wrong with the white suckers that they did catch. Many of these fish were smaller than they should have been. Some were deformed, with flatter than normal heads and curved backbones. Hoping to find out what was happening, Harvey marked these fish with tags and released them back into the water.

The next year Harvey returned. This time he had trouble finding any white suckers at all. He had tagged one hundred fish but was only able to find a few of these. He also could not find any young suckers, even though there should have been new fish born that year. It seemed that the female fish were not producing young.

In an effort to find an answer to the riddle of the dying fish, Harvey took a water sample from Lumsden Lake and tested it for acid. According to government data, in 1961 the lake's pH was 6.8, a very healthy number. Harvey's water sample measured a pH of 4.4. This indicated that the lake was 100 times more acidic than it had been roughly ten years before. Harvey realized that acid rain was affecting even lakes far from the origin of pollution and that this was killing the fish.

Studies uncovered what lakes were acidic and exactly how the acids killed the fish that lived in them. Testing nearby lakes, researchers found many that were acidic. They also found several that, like Lumsden Lake, no longer had fish in them. In order to save the lakes and the fish, scientists needed to understand how acids hurt the fish that live in the water. After much research, science has a clearer idea of how fish are impacted by acid pollution.

The Effect on Fish Populations

Adult fish are more resistant to the lower pH of acidified water, so young fish are the first victims of this pollution. The acids in the water destroy cells, which are microscopic parts of the fish's body. The most important cells affected by the acid are in the fishes' gills, the part of the fishes' body that they use to breathe. When the gills are damaged, the fish have difficulty getting oxygen and other compounds into their bloodstream. The gills of young fish are not well developed and are more easily harmed by the acids. Researchers W. Swenson and T. May demonstrate this point, saying that "adult brook trout can live at pH 3.5 to 4.5, [but] brook trout fry [newborn fish] may die at pH 4.5 to 6.5."[15] The young fish are the first to succumb to asphyxiation, the inability to breathe.

Fish eggs are also extremely susceptible to acidification. When the pH is high, the acidic water burns eggs that are laid in the water, destroying them. According to Swenson and May, "One study showed that in acidified waters, female fish have a higher percentage of eggs that do not develop."[16] In addition, female fish may not lay eggs altogether, because the acid in the water lowers the amount of chemicals in the fishes' bodies that cause them to breed.

When acids in a lake are strong enough to destroy eggs and young fish, only adult fish remain, and eventually they

Fish eggs like these trout eggs are especially susceptible to acid damage. Acidic water burns the eggs and destroys them.

disappear as well. Some of these older fish may have depended on the young as a source of food. Others will die gradually of old age. Since there are no new fish to replace the dying adults, eventually there are no fish remaining in the lake.

Even if acid does not directly kill some fish, it may trigger another type of poisoning that will. Acidification frees up, or mobilizes, toxic heavy metals that naturally occur in rocks and soils, like mercury and aluminum. Once mixed into the water supply, these heavy metals are deadly. Aluminum, even in small amounts, is extremely poisonous to fish. Scientists estimate that one tablespoon of aluminum in a lake the size of a football field can kill all the fish. Aluminum changes the balance of salt and water in the blood, which makes it thicken. According to Ellen Baum of the Clean Air Task Force, aluminum "doubles the consistency of the blood, turning it to the consistency of peanut butter. The fish's heart cannot pump such thick blood."[17] The aluminum basically gives the fish heart attacks.

Effect on Aquatic Plants

Aquatic plants are all affected differently by acid rain. Some plants benefit from the increased acidity, and some plants suffer. However, any change to plant life affects all other organisms that live in the lake.

Alga, a very simple plant, is one of the first plants to be affected by a change in acidity. The dominant type of algae and other simple plants may die off rapidly in response to a rise in pH. This is a problem in a lake ecosystem because algae are the primary source of food for zooplankton, the microscopic animals that some species of fish depend on as a food source. If the algae disappear, then the zooplankton dies off. Once the zooplankton has disappeared, the fish lose their food source and die out as well.

There are several species of algae, however, that thrive in acidic water. These species of thick green algae replace the species of algae that were food for the zooplankton. These algae, such as the genera *Mougeotia*, grow into thick stringy mats of green on the bottom of acidified lakes. The algal mats can cover other aquatic plants, blocking the sunlight. Without sunlight, other aquatic plants die. The destructive algae also reduce the available fish-nesting sites and hiding places created by other plant life at the bottom of the lake.

Sometimes acidification can cause an increase in the amount of nitrogen in lake water. Nitrogen is a nutrient, but in increased amounts it can still damage lakes. Nitrogen is one of the elements in acid rain. When the concentration of this nutrient in a lake rises due to acid buildup, more plant life is able to grow. With more nutrients available, more and more plants grow until there are many more plants than the ecosystem can maintain. Eventually some of these plants die naturally, and a large amount of decaying plant matter is left in the lake. The process of decomposition uses up all the oxygen in the water and suffocates everything else living there. According to John McCormick, "the water becomes virtually lifeless and foul-smelling."[18]

Indirect Effects on Animals

As a lake begins to die, some other wildlife may be affected indirectly. Animals that do not spend all their time in the water, such as birds, otters, and minks, may not be hurt by the rising acidity. However, as the fish that they depend on for food disappear, they will suffer as well.

Aquatic birds such as the loon, the merganser, and the belted kingfisher have to change their food of choice or find other lakes when fish in an acidified lake begin to disappear. For example, the common loons have a diet of 80 percent fish. If there are no fish to eat, the birds must find another source. Soon there are no birds left on the dead lake either.

In Norway, studies have shown that the dipper, Norway's national bird, has been affected in this manner by water acidification, and scientists are worried that it may become endangered. The birds eat mainly animals with shells, like mussels, which are one of the first animals to disappear as water becomes acidified. Research has shown that these shelled animals are a crucial part of the bird's diet.

A dipper perches on a rock. Acid rain is responsible for significantly reducing populations of shellfish on which the dipper depends for food.

Many aquatic birds depend on the shells of snails, mussels, and crayfish for calcium in their diet. Calcium is used in developing a bird's eggshells. Without the added calcium of clams, the eggshells will be very thin and break before the chicks can develop. If there are no new hatchlings, the birds will eventually disappear. Scientists have been monitoring Norway's dipper since 1974, and problems with its thin-shelled eggs have increased since studies began.

Since the 1960s, scientists and citizens in many countries have become increasingly aware of the far-reaching effects of acid rain and acidified lakes. Once a lake becomes acidified, all the animals that are associated with the lake are adversely affected. Dead and dying lakes are the most obvious result of acid rain. As researchers have examined the problems and possible solutions to dying lakes, they have recently discovered that acid rain is harming other less obvious ecosystems as well.

3

Dying Forests

ONE OF THE most beautiful shows of nature in the United States is the changing colors of the leaves in the autumn. The trees in the New England and mid-Atlantic states put on such a spectacular display that tourists come from all over the country to drive along the country roads, enjoying and photographing the deep red, orange, and yellow colors of the leaves.

The changing and falling of the leaves is a natural process. As the days grow colder and shorter, trees begin to prepare for the winter months. The leaves on the trees are food-making factories, and to conserve energy, the trees begin to shut them down. During the winter there is not enough light or water for the leaves to create food. Instead the trees will live off the food that they stored up during the summer. As these leaves die, they change color and fall to the ground.

Concern for Pennsylvania's Forests

In recent years, however, tourists traveling across Pennsylvania in autumn, searching for autumn colors, have been surprised at what they have seen, or rather, what they have not seen. In some places the colors are completely missing. Instead, what tourists are finding is large sections of forest with only the brown stumps of dead, leafless trees. Many bright leaves and healthy trees have disappeared along the state's scenic hillsides.

Worried researchers at Pennsylvania State University have investigated this phenomenon and have come to a conclusion: Acid rain is killing the Pennsylvania forests. According to Bill Sharpe of Pennsylvania State University's School of Forest Resources and Environmental Resources Research Institute, "Hundreds of thousands of acres are affected by acid rain here in Pennsylvania, with the trees that give us the bright fall foliage among the hardest hit. Unless we get the acid rain problem under control, we will lose more and more trees."[19]

Among the hardest hit species of these trees is one of great historical and economic importance, the sugar maple. The sugar maple is the state tree of New York. The sugar maple is also important to Canada, and the maple leaf is even on Canada's national flag. Sugar maple sap has been used to make syrup since the early 1600s. The tree is also commonly used for its hardwood timber. Today, researchers believe that the maple is in trouble in Pennsylvania and potentially in New England and Canada as well.

A stand of maple trees in Susquehannock State Forest in Potter County, Pennsylvania, has been made an example of the deadly consequences of acid rain. Sharpe states, "These trees will not have normal fall color because the leaves are already stressed by the loss of necessary nutrients. The acid rain has pulled [nutrients] from the soil, so the leaves are very small, and many are turning brown and dying. The leaves actually began dying early in the summer, so they won't develop full fall color as a result. In this particular area, about 90% of the sugar maples are dead or dying from acid rain."[20]

In the early 1980s, scientists felt that acid rain was not a threat to American forests. Some researchers even thought that the nitrogen in acid rain was beneficial to forests, fertilizing them and helping them grow. Today most scientists believe that acid rain damages forests. Trees with damaged or missing leaves are unable to produce food effectively through photosynthesis. The trees, therefore, die slowly. The sugar maple is not the only tree affected by this tragedy, nor is the United States the only country to witness the decline of forests through acid rain.

The autumn colors of sugar maples like this one have dulled in recent years, a phenomenon researchers attribute to leaf damage caused by acid rain.

A Worldwide Problem

Scientists have been in agreement since the 1980s that the forests of the world are in trouble, but it was not until the late 1990s that studies demonstrated that the problem in Pennsylvania was acid rain. Today many researchers still argue over the exact damage done by the acidic rain and how it occurs. There is no arguing, however, that in many places, including Norway, Poland, and Germany, as well as the United States, parts of the forests are dying.

The forests of Pennsylvania are the most notable site of the occurrence of acid rain in America, but the alarm call was first made in Germany and the other forests of western Europe. When the silver fir trees began to brown and die after a drought in 1976, foresters, people who make a living off the forests, were concerned. Most assumed, however, that the forests would quickly recover. In the years following, it became apparent that the trees were not returning and more were dying off.

German's take pride in their silver fir trees, the same way American's take pride in their sugar maples. The popular Christmas carol "Oh, Tannenbaum" celebrates the beauty of the silver fir tree. The fir tree is often mentioned in German folk stories and is also used to decorate homes during the holidays. Ecologist Chris Rose notes, "It was not the foresters who first noticed that the forests of central Europe were dying, but the parents and children who went each year to the Black Forest in West Germany to collect silver-fir foliage for Christmas decorations."[21] Foresters reported that people saw how discolored the trees were and stopped coming to buy fir trimmings for their homes.

The German people were unsure what was happening to their trees, but they were aware that the forests were in trouble. Death by drought, insects, and other known tree diseases were ruled out as a cause. Scientists were puzzled as the death spread to more species of trees. It was then that the people gave this mysterious destruction a name. They called it "Waldsterben," which means "forest death" in German.

Waldsterben

Waldsterben had entirely different symptoms from the normal causes of forest decline, such as diseases from fungi, bacteria, and viruses. In conifers, trees like the silver firs that have needles, the trees' needles turned yellow and fell to the ground. These trees also had stunted growth, thinning of treetops, soft root rot, and root damage. In deciduous trees, like oak trees that have leaves, scientists found discolored and misshapen leaves. Their leaves fell early, treetops were dying, and bark was damaged. The deciduous trees seemed less tolerant to other stresses, like cold and insects, as well. All of these trees had stunted growth, and in the worst cases they died. Scientists pinned the blame on acid rain caused by the industrial areas of Germany and other nearby European countries.

As Waldsterben continued, forests were affected in more European countries. According to author John McCormick, "By 1985 almost every species of forest tree in Europe was affected, with damage evident on more than five million hectares (12 million acres) of forest in just four European countries: West Germany, Poland, Yugoslavia, and Czechoslovakia."[22] Waldsterben became so widespread that Europeans realized that it had to be monitored across the continent. In order to keep track of the damage, a Europe-wide program to monitor forest health began in 1987.

Today Waldsterben is still destroying trees in Europe. The Federal Research Centre for Forestry and Forest Products reported that in 2001, 22.4 percent of the 132,000 trees

Spruce trees in Poland show the devastation of Waldsterben. This phenomenon destroys thousands of European trees every year.

assessed were classified as damaged. Trees are considered damaged when they have lost 25 percent of their leaves or needles. The 2001 study shows an improvement since 1995 when forests were at their worst, with 25.8 percent of the monitored trees damaged. Scientists point out, however, that there has been a slow and steady increase in damage over the last several years.

Forests at Risk

Although many forests are currently suffering from the effects of acid pollution, some forests are more susceptible to acid damage than others. Researchers have studied damaged and healthy forests to evaluate what the risk factors are. In this way they hope to be able to determine what forests are in the most danger.

One of the largest factors in determining the susceptibility of a forest to acidification is its soil. Just like lakes, forest soils that are rich in limestone or other buffering agents take longer to become acidified or may not become acidified at all. Soils that run deep are also less susceptible to acidification than areas of shallow topsoil. Thus, the soil type and depth help indicate how at risk a forest is from the damages of acid rain.

The features of the forest, such as the types of trees and the variety of the landscape, also help shape a forest's susceptibility to acidification. Some tree species are more likely to be harmed by acids than others. If a forest is made up of acid-sensitive species, such as sugar maple, the forest is more likely to decay. Landscape features such as elevation and steepness of the hills also affect acidification. High-elevation forests are more likely to receive acid rain due to the cool, wet climate. Forests on a slope are less likely to become acidified than the forests in the valleys below because rainwater drains into the lowlands. Scientists look carefully at forest features when examining the effects of acid rain.

Finally, the weather that a forest receives greatly affects its risk of acidification. The greater the amount of rainfall a forest receives, the more acids will be added to the environment.

Wind also plays an important role. Forests that are downwind from polluting factories, even if they are many miles away, are more likely to receive acid precipitation.

Acid Fog

Fog can also contribute to acidification. High-elevation forests are foggy much of the year. This fog can contain acid, just as in the acid rain that falls at lower elevations. The longer the fog sits on the forests, the more likely the forests are to become acidified.

Mount Mitchell is an example of a forest region suffering from acid fog. Mount Mitchell is located in the Blue Ridge Mountains and is the highest point in North Carolina. The mountain is so high that it pokes through the clouds and is almost constantly bathed in fog. In fact, the mountain is covered in fog eight out of every ten days. This is a problem, because acids are concentrated in the clouds. The clouds have a much lower pH (higher acidity) than the rain that falls from them. The lowest recorded pH at Mount Mitchell was 2.1, about the same acidity of bottled lemon juice. Today the peak of Mount Mitchell is covered by dead and dying trees.

Direct Effects of Acid Rain

Any form of acid precipitation will have a direct effect on the trees and plants of a forest. Acids can burn through the protective layer on leaves. Leaves have a coating on the outside that feels waxy to the touch, but this coating is quite thin. When acids are deposited on the leaves and remain there, they burn through the leaves' protective coating. The burning produces brown spots on the leaves. If enough burned areas occur, the leaves will typically die.

Pine needles are also damaged by acid rain. Pine needles stay on conifer trees all year long, even in the middle of winter when leafy deciduous trees have lost all their leaves because of the cold. Pine needles are designed to withstand the cold, but acid rain reduces their endurance.

Pine needles have a high calcium content. In conifer trees, calcium stops needles from freezing when temperatures get

Acid rain has stripped these fir trees of their needles. Acid rain leaches calcium from the needles and makes them unable to withstand cold weather.

low, but calcium is also a buffering agent that helps neutralize acid. In order to neutralize it, though, it has to combine with the acid. In this way acid rain ends up leaching the calcium from the needles. Once the calcium has been completely removed, the conifer needles freeze and die. Since the tree has a lowered resistance to freezing damage, it often perishes in the cold.

Direct damage to conifer needles is killing red spruce in the United States. According to researcher Ellen Baum, "Since the 1960s, more than half the large canopy red spruce have died in the Adirondack mountain of New York and Green mountains of Vermont, and roughly 25 percent have died in the White Mountain."[23]

Death of established trees can make it difficult for seedlings to survive as well. Large trees make up a canopy above the smaller trees of the forest. These trees filter the sunlight that reaches forest floor. As the tall trees die, the amount of light that reaches the forest floor changes. Plants that thrive in this sunlight begin to grow rapidly and choke out the young tree seedlings that are growing on the forest floor.

When large trees die within the stand of forest trees, it is not only the young trees that are in danger; survival can become difficult for the entire forest of trees. When all the trees grow closely together, they have more protection from the weather, especially the wind. When trees die within the forest, the remaining forest is fragmented and does not have as much protection. A heavy windstorm can blow them down, causing increased tree falls. Acid rain, therefore, has a devastating effect on the trees it touches, but it also has an indirect influence upon the health of the forest by poisoning the soil of the forest floor.

Effects on Soil

Researchers agree that the majority of dying forests are suffering from changes in the soil caused by acid rain. Until the 1980s, there were no significant studies of forest soils in the United States. Researchers have spent the last twenty years engaged in studies of forests where the threat of acid rain is known. Today scientists have a better understanding of the three primary changes acid rain causes in forest soil.

Acid rain causes a depletion of calcium in the soil, and scientists believe that this is detrimental to all trees. Just as in the leaching of calcium from conifer needles, acids from the rain bind with the calcium and remove it from the soil. Calcium is essential to wood formation in trees, so trees have a very high demand for calcium. When calcium becomes unavailable, the tree is unable to grow at a normal rate.

The slow growth rate affects both young and old trees. Older trees may be unable to recover from insect damage, since the trees are missing an essential nutrient for repair. Seedlings are simply unable to get calcium from the soil necessary to reach full growth. In Ontario and Quebec, this nutrient loss has caused a decline in sugar maple seedlings. Without new trees to replace the dying ones, Canada's sugar maple forests are diminishing.

Another change in soils caused by the introduction of acid rain is the mobilization, or release, of aluminum that is normally trapped in the earth. When enough aluminum is active in the soil, it can become toxic. The aluminum damages the fine roots of the trees. This prevents the tree from getting enough nutrients and water to survive. Once

A Canadian scientist collects soil samples for research. Soil contaminated by acid rain has proven deadly to many of the world's forests.

the roots are damaged, the rest of the tree suffers and may die. Studies have shown that the tulip poplar, a major tree variety in southern Appalachian forests, has been subjected to toxic levels of aluminum in this manner.

The most apparent change in soils from acid rainfall is the addition of sulfur and nitrogen. The sulfuric and nitrous acids from the rain accumulate in the soils and raise the pH. As the pH of the acids changes in the soils, species of plants and trees may have difficulty surviving. Most species of plants and animals can only survive in a narrow range of the pH scale. If the soil is too acidic, the trees will not grow as well.

In most situations where trees are dying from acid rain, the problem is a combination of all of these soil changes as well as the natural stresses in the environment. Trees that are weakened by the lack of calcium or an inability to retrieve water and nutrients from the soil will have a harder time recovering from normal diseases and parasites that would be unable to kill a healthy tree. And once trees start to die, the entire forest ecosystem is in danger.

Killing the Forest Ecosystem

A forest ecosystem involves much more than its trees. There are many insects, animals, and plants that depend on the forest for survival. Once acid rain begins to affect the soil, the very foundation of the forest, much of the plant life and wildlife will soon be in jeopardy.

As the pH in the soil falls, bacteria, fungi, and earthworms may die. Bacteria and fungi live on the forest floor and are important because they feed on decaying plant matter and animal droppings that litter the ground. They break down nutrients in this debris and add it to the soil. Under the ground, earthworms break up leaf litter into pieces so tiny that the trees absorb the nutrients through their roots. Earthworms are unable to survive at a pH below 4.0 and will perish in acidified soil. Without these organisms to break down the forest debris, more and more material builds up on the ground.

When the debris on the forest floor gets too thick, the forest cannot rejuvenate itself. Tiny seedlings that begin to grow at the top of the leaf litter cannot work their way down to the soil to take root. The cycle of growth and decay that would sustain the forest indefinitely is disrupted, and the decline of the forest is inevitable.

The poisoning of forests impacts animals that feed on the creatures that live in the soil. A recent study from Cornell University proves that wood thrushes, which thrive in the eastern United States and Canada, are less likely to breed in areas damaged by acid rain. According to writer Robert Winkler, "In areas where acid rain is most severe, the supplementary calcium-rich foods that female songbirds depend on, snail shells, isopods, such as pill bugs, millipedes, and earthworms may be in short supply."[24] In forests with high soil acidity, these snails and isopods disappear and are not available for food. The female songbirds are then more likely to lay thin and brittle eggs that break during incubation. If the eggs survive and hatch, the parents will have a difficult time finding food that meets the high calcium needs of the young birds. Today there are fewer and fewer wood thrushes breeding in these acid-stressed forests.

The damage done to the forest also harms animals that live there. Trees provide nesting sites and protection for

As trees die from acid pollution, birds like this wood thrush are deprived of nesting sites and sources of food.

animals like squirrels and woodpeckers. When these trees disappear, the animals have difficulty breeding and escaping predators. During the winter, the absence of trees also allows more snow to reach the ground. The deep snow makes it harder for animals like rabbits and deer, to find food. With fewer trees and shrubs, there is also less protection from the cold. The animals that make a home in dying forests have a difficult time surviving.

Once the acid rain begins to damage trees in the forest ecosystem, a cycle begins that hurts all of the animals and plants that depend on the forest for survival. This process may start small but quickly becomes a catastrophe. Since the process of forest death involves many factors, scientists still feel there is much to learn about what is killing forests. Most scientists agree, however, that acid rain is the start of this destruction.

4

How Acid Rain Affects Humans

IN THE EARLY 1900s, London had long been known for its smoky, polluted air. The smoke of burning coal mixed with the fog that often fell on London, turning the air dark, murky, and dangerous to breathe. On December 5, 1952, this London "smog" caused the deadliest environmental catastrophe in history.

A funeral director named Stan Cribb witnessed the beginning of the "Killer Fog of '52" and remembers it clearly even fifty years later. He has led thousands of funeral trains through the smoggy streets of London, but he says that 1952 was the worst smog he had ever seen. Describing the first gray wisps of the smog, he says, "You had this swirling like somebody had set a load of car tires on fire."[25]

On Friday, December 5, Stan Cribb and his elderly uncle were driving to a wake, with a line full of mourners behind them. As they were driving, the smog became thicker and blacker. After a few minutes, Cribb could not even see the curb as he was driving the hearse down the road. His uncle got out and tried to guide the hearse by walking in front of it, carrying a powerful hurricane lantern in one hand. It was useless. Cribb states, "It's like you were blind."[26] They had to stop the funeral train, and they began to wait, but it would be four days before the smog blew away.

The Killer Smog

The day before the smog began had been cold and windy, which was the start of the problem. Londoners trying to keep warm burned large quantities of coal in the open grates of their homes. The coal they were burning was inexpensive, which also meant that it created a lot of smoke. The good coal was being exported out of the country by the government, which was close to bankruptcy and trying to generate money. The inexpensive smoke-producing coal sent a tremendous amount of smoke into the London air. When the fog came the next day and the wind disappeared, the smoke became smog and settled on the city.

On the second day of the smog, five hundred people died in London and thousands of others became ill. The ambulances had stopped running because it was impossible to see well enough to drive a vehicle. People who were sick had to walk through the smoggy streets to the city hospitals. Many arrived at the hospital unable to breath.

Many Londoners already had weak lungs from living in the polluted air or from smoking. Damage to their lungs meant that their lungs could not hold as much air as a person who had healthy lungs. These people would appear to have normal, healthy lives, but if they participated in strenuous activity, like climbing stairs, they would have difficulty breathing. When the air quality diminished, they would be unable to breathe and were the first to die. They would breathe deeply but could not get enough oxygen.

Acids in the air were often the final blow to the sick. Many people came into the hospitals gasping for air as their lips turned blue because they could not get enough oxygen to their blood from the polluted air. As they breathed in this air, acids were sucked into their lungs as well. These acids triggered massive inflammations, meaning the lining of the lungs became irritated and swollen. The swollen tissue could not absorb oxygen, and the people suffocated.

On Tuesday, December 9, the wind picked up and swept the fog away, returning the city to normal, and it seems that many chose to forget that it ever happened. This event is not

London traffic is at a standstill during the "killer fog of '52." In December 1952, a thick, acid-polluted fog settled over the city and killed nearly twelve thousand people.

as well remembered in the United States as the killer smog of Donora, Pennsylvania, four years earlier. However, the acid-polluted fog of 1952 is thought to have been responsible for twelve thousand deaths in London, the largest death toll of any human-caused environmental disaster.

Incidents like the acid fogs in London and Donora are not common. It is unusual for such large amounts of pollution to get trapped close to the ground, forcing people to breath poisonous air. However, in these instances it is easy to see how acids in the air can harm human beings. Scientists point out that exposure to even small amounts of acids in the air can harm human health over time.

The Dangers of Acids in the Air

The two main ingredients of acid rain, snow and fog, are nitrogen oxides and sulfur dioxide. These gases become dangerous when they are mixed with water because they become acids. If the acids are mixed in rain or snow, they are not likely to directly affect human beings. However, if they are mixed in fog or remain in the air as tiny particles, humans can breathe the acids into their lungs.

The air is full of particles that humans might breathe into their lungs. Dust, pollen, and other particles occur naturally. The human body has evolved defenses to keep too many of these particles from getting inside the body.

There are several ways that the human body stops particles from being inhaled into the lungs. The hairs on the nasal passages trap many of the larger particles, such as dust. There is also mucus in the nose where these particles can get caught. If a person is breathing through his or her mouth, particles may not pass through the nose. If this happens, small hairs in the lungs called cilia wave back and forth and carry the particles out.

However, if particles are small enough, they can get past the body's defense system. This is what makes air pollution so dangerous. The particles that humans create by burning fossil fuels are much smaller than dust or pollen. Sulfur dioxide and nitrogen oxides are especially harmful because they are such small particles. These particles are drawn straight into the lungs where they can damage the delicate tissue.

Nitrogen dioxide is a deep lung irritant. According to the Minnesota Pollution Control Agency, even "short term exposure (e.g., less than three hours) to current nitrogen dioxide concentrations [in Minnesota] may lead to changes in airway responsiveness and lung function in individuals with pre-existing respiratory illnesses and increases in respiratory illnesses in children."[27] Long-term exposure can cause a buildup of fluid in the lungs, which in severe cases can cause death. It can also weaken the tissue of lungs, which makes it easier for the lungs to become infected.

Sulfur dioxide irritates the whole respiratory system. This irritation leads to constriction of the airways, which makes it difficult for oxygen to get to the blood. This is especially common in individuals with asthma. Asthmatics exposed to sulfur dioxide may experience wheezing, chest tightness, and shortness of breath. Long-term exposure to sulfur dioxides can also result in respiratory illness, like colds and pneumonia, and can even cause an existing heart condition to worsen.

Acid pollution in the air can also cause minor problems that people might not even notice. Although the people that air pollution impacts the most are young children, the elderly, and those that are already fighting a respiratory disease or infection, perfectly healthy people also suffer. These people may simply suffer from a dry cough or throat irritation or have difficulty getting over a current cold. Doctors agree, however, that the biggest problem acid pollutants cause involves people who suffer from asthma.

The Asthma Epidemic

Asthma is a disease that causes the airways in the lungs to be constantly inflamed. When this happens, the airways become constricted, or closed, and it is difficult for a person with asthma to get oxygen. Since asthmatics have airways that are constantly constricted, they often have attacks of wheezing or shortness of breath.

Asthma affects 17.3 million people in the United States and results in more than twelve thousand asthma-related deaths each year. The numbers of people who have asthma continue to rise every year, and this has doctors concerned. Their largest concern is for children, who have the most difficulty with the disease.

In 1998 the President's Task Force on Environmental Safety Risk to Children declared asthma a national epidemic. Dr. Philip Landrigan, director of the Center for Children's Health and the Environment of New York, says, "Despite advances in therapy, asthma attack rates among American children have more than doubled in the past decade. Death rates are also rising."[28] Scientists are examining the effects of acid pollution on children and looking for solutions.

Children are more likely to be harmed by air pollution than adults. Children breathe in twice as much air as adults, and this makes them extremely vulnerable to acid pollution. Children also have smaller respiratory airways, which means that they are more sensitive to irritants, like acids, than adults are. Children are also more likely to play vigorously outdoors, causing them to breathe in more pollutants than adults. Scientists feel that all of these factors are contributing to the asthma epidemic, the effects of which are apparent. According to a Massachusetts state Senate report. "On average, school children across the country will miss 14 million school days a year due to health problems related to asthma."[29]

Doctors are especially worried about children who live in urban areas where there is more acid pollution from cars and industrial smokestacks. Studies have shown that children from these areas suffer a higher rate of asthma than children from the more rural areas of the country. Massachusetts state senator Dianne Wilkerson was surprised by the number of children with asthma in a school in Roxbury, a neighborhood in Boston. At an after-school program, she noted that there were asthma inhalers in seventeen out of

A young asthmatic breathes with the help of an inhaler. Many doctors blame air pollution for the rise in recent years of asthma cases.

twenty cubbyholes. Wilkerson was shocked and stated, "That's not the way it used to be and that's not the way it's supposed to be."[30] In Roxbury the rate of asthma-related hospitalizations among children is five times higher than the state average.

Asthma is more than a nuisance; children can die from the disease. Asthma can strike at any time, causing a child to be unable to breathe. The child gasps for air and is unable to get enough oxygen, which causes their oxygen-poor blood to turn their lips and fingernails blue. As breathing becomes more difficult, the heart has to work harder than normal. If the airways do not loosen and let in air, the child can suffocate or have a heart attack.

The Massachusetts Senate report points to one such incident. In the middle of the night in March 2001, one child had an asthma attack at home. Sixteen-year-old Marquisse McGregor stopped breathing as a result of this attack. Emergency workers tried to revive him, pumping air into his lungs, but they were unsuccessful. The boy, a Brighton High School student, died.

Aggravated asthma and respiratory problems are direct effects of acid rain, but humans suffer from indirect effects as well. The direct effects of acid pollution are caused by acids that are mixed in fog or by the precursors of acid rain, sulfur dioxide and nitrous oxides. Once acids fall to Earth as rain or snow, they have little direct effect on humans. To people, acid rain feels, and even tastes, like normal rain. However, the problems that this rain causes in the environment can also harm humans.

Contaminated Food

One of the ways in which acid rain causes harm to humans indirectly is by releasing heavy metals, like lead, mercury, and aluminum, into the ecosystem. Heavy metals are commonly found in all environments, but most of the time they are harmless because they are bonded to other metals. Acids unglue these metals from rocks or soils. Once separated, the metals may be carried by rain runoff into a lake or other water source.

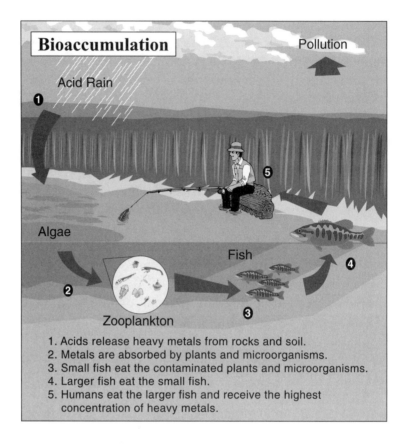

1. Acids release heavy metals from rocks and soil.
2. Metals are absorbed by plants and microorganisms.
3. Small fish eat the contaminated plants and microorganisms.
4. Larger fish eat the small fish.
5. Humans eat the larger fish and receive the highest concentration of heavy metals.

Once metals are in the water, they are absorbed by the plants and microorganisms. These plants and microscopic animals have only small amounts of the metals within them. However, small fish consume a large amount of these plants and animals, and the concentration of the metals in the small fish is higher. Bigger fish eat a large amount of small fish and then have an even higher concentration of these metals that they are unable to excrete from their bodies. The metals then work their way up the food chain. This process is called bioaccumulation. According to researchers T. May and L. Klessig, "animals higher on the food chain may have thousands of times the levels of toxins as those concentrations in the water itself."[31]

When heavy metals get into the body of an animal, they do not leave. In small amounts this is not a problem, but if an animal or human continues to accumulate metals in the body, it may cause health problems over a lifetime.

Aluminum

Aluminum is one of the harmful metals released by acid rain. Aluminum is the most abundant metal on Earth. Since it is such a common metal, it is almost always found in acidified soils and lakes. The fish in acidified lakes often have high levels of aluminum in their bodies.

In the past, scientists believed that aluminum was not toxic to humans because it is so abundant in the environment. Recent studies have proven that this is not true. According to Health Canada, the federal department responsible for Canadian healthcare, "abundant evidence now shows that aluminum may adversely affect the nervous system in humans and animals."[32] These studies show that if humans eat a large number of fish from acidified lakes, they may accumulate dangerous levels of the metal in their bodies.

Many scientists believe that aluminum exposure may be the cause of Alzheimer's disease. This disease begins with mental deterioration, such as memory lapses, confusion, and depression. As the disease progresses, the individual becomes more confused and over time is no longer able to care for him or herself. Eventually, the victim's brain function deteriorates to a point where the body cannot function, and the victim usually dies.

Scientists are unsure exactly how the disease is caused. Research has shown, however, that the brain cells of Alzheimer's patients can contain ten to thirty times the normal concentrations of aluminum. This fact has led many scientists to believe that aluminum may at least in part be a cause of the disease.

Aluminum has also been associated with other severe diseases of the nervous system. Lou Gehrig's disease and Parkinson's disease are similar to Alzheimer's, causing gradual deterioration of brain function. Sufferers of these two diseases also have high levels of aluminum in parts of the brain.

Although scientists are not completely certain that aluminum is the cause of Lou Gehrig's and Parkinson's

An Alzheimer patient works with a physical therapist. Some doctors believe that increased exposure to aluminum, due to acid rain pollution, may be a cause of Alzheimer disease.

diseases, a study in Guam and New Guinea shows that it is certainly a possibility. The local environmental conditions in the study area of these two countries included high levels of aluminum in the soil and food. In these areas there is an unusually high rate of Lou Gehrig's and Parkinson's diseases in the native population.

High levels of aluminum can also have other effects on the human body. Aluminum can cause anemia, a condition in which the body processes fewer red blood cells, making the sufferer feel tired and weak. Doctors say that aluminum can also cause brittle or soft bones. Aluminum at toxic levels has been linked to heart attacks. Overall, scientists agree that accumulation of large quantities of aluminum is dangerous to human health.

Mercury

Mercury is the most toxic of all natural metals. Although the recognition of poisoning from aluminum is new, mercury poisoning has been a known toxin for hundreds of years. Mercury is commonly used by dentists, fingerprint photographers, and hospital technicians, and it was once used by miners to extract gold from stream gravel. All of these professionals have to be extremely careful, because a small amount of mercury can be deadly.

Many people are familiar with the Mad Hatter from Lewis Carroll's *Alice's Adventures in Wonderland.* The character was modeled after the hatters of the 1800s who designed fancy hats for high-class women and men to wear. Many of these hats were made from felt, and mercury-infused water was used in order to shape the felt. Hatters stood in this poisoned water, shaping felt for as long as ten hours a day. These hatters often developed the shakes, slurred speech, and manic-depressive behavior. Many of the hatters died, and the ones who did not were often classified as "mad" and were sent to mental institutions.

Today people are much more careful with mercury, but as the metal accumulates in fish from acidified lakes, people can still get poisoned. Today doctors are noting mercury poisoning in people who consume large quantities of fish. For example in 2002, Suzie Piallat was tired and achy and could not concentrate, so she went to see Dr. Jane Hightower in San Francisco. According to *Associated Press* reporter Sharon L. Crenson, "When Hightower asked Piallat if she ate a lot of fish, she said yes: eight meals a week. And when Hightower tested her blood, she found mercury levels of 76 parts per billion, 15 times the amount considered safe by the Federal Centers for Disease Control and Prevention."[33]

Although adults suffer from mercury poisoning through fish consumption, children are even more vulnerable. The latest Food and Drug Administration (FDA) guidelines recommend that pregnant women and small children eat no more than two meals of fish a week. These FDA recommendations are based on a study published in 2003 that took place in the Faroe Islands, a group of islands between Norway and Iceland.

The University of Boston conducted the study in the islands and discovered problems with children born from mothers who were contaminated with mercury.

The islanders consume whale meat and blubber. Although not a fish, whales take in a high amount of the ocean's mercury through the plankton they eat. The high levels of mercury are believed to be the result of acid pollution. The women that had high levels of mercury in their systems gave birth to children with poor language skills, memory, and attention spans. Dr. Phillip Landrigan, chairman of the Mount Sinai medical school's Department of Community and Preventative Medicine, says of mercury, "It's the chemical that can push a child over the edge"[34] into a lifetime of health problems.

The potential danger of mercury poisoning is indisputable, and many scientists are worried. Mercury warnings for U.S. rivers, lakes, and coastal regions have increased 115 percent from 1993 to 2001. There are almost two thousand mercury-in-fish warnings on different water bodies in forty-four states. Today, people who live near lakes and streams that are affected by acid rain are warned to be careful about the amount of fish that they eat.

Acid in Drinking Water

While eating fish and other seafood can pose a danger, humans are less likely to be poisoned from water contaminated by mercury and aluminum. For the most part, the

A sign warns of the health hazard of eating fish contaminated with mercury. Fish that live in acidified lakes often absorb toxic levels of mercury and can poison people who eat them.

levels in the water are much too low to affect people. However, scientists are still uncertain how a lifetime of accumulated heavy metals might affect humans. There is still much research to be done in this area.

The one way acid rain does immediately damage drinking water is by pipe corrosion. Just as in the environment, acids can loosen metal from water system pipes and release them into the drinking water. Today most pipes are plastic, but older homes have copper pipes that, in combination with acid rain, can be dangerous.

Copper and lead can be released from pipes and poison humans when they drink the water. If acid water sits in pipes for a few days (if the family of the house goes on vacation, for example), the acids will release metal particles into the water. Copper pipes are bound together with lead and tin. Lead is extremely dangerous and can harm the human brain and nervous system. The copper of the pipe itself is dangerous as well.

People can avoid poisoning from old pipe systems by running water for a few minutes to flush out the system. However, there is no arguing that acid rain can affect humans both directly and indirectly. To ensure the safety of their citizens, researchers in countries around the world are investigating ways to end the acid rain problem.

5

Searching for Solutions

TODAY SCIENTISTS, POLITICIANS, and citizens agree that acid rain is a problem that must be solved. However, the solution is much more complicated than the decision to act. Acid rain is created by things that humans use on a daily basis, like electricity and vehicles. Few people are willing to live without electricity and transportation. Still the pollution that is created by daily living must be reduced to lessen the problem of acid rain. The dilemma between a demand for inexpensive electricity and transportation and a need for cleaner air prompted a debate between Canada and the United States that demonstrated the complexity of agreeing on solutions to acid rain.

The Debate Between Canada and the United States

In the early 1980s, the United States was becoming increasingly aware of the environmental problems that were possibly being caused by acid rain, but the government was reluctant to pass legislation to alleviate acid pollution. Then current president, Ronald Reagan, had pledged to reduce federal legislation during his presidency. President Reagan felt that too many policy mistakes had been made in the past due to a lack of scientific information. He did not want to pass any new laws without good reason, and because of this, he was opposed to supporting anything other than further research of acid rain.

This approach to the problem of acid rain did not please the Canadian government that was accumulating research proving that a large amount of Canadian pollution was originating in the United States. The damage from wind-borne pollutants originating in America was much greater in Canada than in the United States. Not willing to ask the United States to do what the Canadians themselves would not do, the Canadian government in 1982 offered to halve their sulfur dioxide emissions if the United States agreed to do the same. Still, the Reagan administration would only agree to more research. According to John McCormick, "in March 1984 Canadian government officials were quoted as saying that US inaction on acid rain was one of the biggest irritants in US-Canadian relations."[35]

Discussions between the two countries continued, but the Reagan administration was reluctant to agree to any new laws. The technology required to reduce acid pollution would cost power plants money, and that would mean higher electricity prices. The president did not wish to pass laws that would cost the American people money and cause difficulties for the utilities industry, especially since the government was uncertain to what degree acid rain was harming citizens. The U.S. government was more concerned about other pollution crises that had a more pronounced effect on humans. Toxic and nuclear wastes were given the highest priority in policy decisions. The wait-and-see approach to the acid rain dilemma continued to raise tensions between the two countries.

In the late 1980s, the studies President Reagan had funded highlighted the amount of damage acid rain was doing to forests, lakes, and even humans. With this increasing amount of evidence, U.S. policy makers realized that Canada was correct; the damage caused by acid rain needed to be confronted. According to writer John McCormick, "a 1987 report by the EPA [Environmental Protection Agency] suggested that perhaps the agency was concentrating on the wrong problems."[36] Many scientists felt that toxic and nuclear wastes were important issues but not more important

Former president Ronald Reagan speaks at a press conference. Reagan was reluctant to pass legislation to reduce acid pollution, but he did support research into the effects of acid rain.

than acid rain. However, President Reagan did little to alleviate the problem, and the tension between the United States and Canada carried through the end of his term in 1989.

In the 1988 U.S. presidential campaign, George Bush promised to address the issue of acid rain, understanding that the public was becoming increasingly concerned about acid pollution and relations with Canada. When President Bush was elected, he followed through with this promise and in June 1989 proposed that major revisions be made to the Clean Air Act. These changes satisfied both Canadian environmentalists and the government, and relations between Canada and the United States returned to normal.

The Clean Air Act Amendments of 1990

The president wished to focus on what he felt were the three major threats to the nation's environment and the health of the Amercan people. He proposed that the Clean Air Act focus on solving the problems of acid rain, urban air pollution, and toxic air emissions. Congress worked on the proposal, approved it, and returned it to the president, who signed it, making it law on November 15, 1990.

Congress stated in the amendments that "the presence of acidic compounds and their precursors in the atmosphere and in deposition from the atmosphere represents a threat to natural resources, ecosystems, materials, visibility, and public health."[37] The act proposed several goals to combat this problem. In 1990 it was estimated that 20 million tons of sulfur dioxide were emitted annually by the United States, mainly by power plants. The Clean Air Act called for a reduction of 10 million tons of emissions by the year 2010. The Clean Air Act also sought to lower the emissions from vehicles by creating tighter pollution standards for emissions from automobiles and trucks, which are the main sources of nitrous oxides.

Pollution Allowances

The Clean Air Act works to reduce the emissions from power plants by an emissions allowance program. The Environmental Protection Agency issues a certain numerical "allowance" to each power plant in the nation. The plant's allowances work like currency that can be spent to release a like number of tons of sulfur dioxide from a smokestack. The power plant can only release as much sulfur dioxide as it has allowances for. If a plant expects that it is going to release more sulfur dioxide than it has allowances for, it must get more allowances. The plant can do this by buying allowances from other plants that have produced less pollution and, thus, have allowances to spare. In this way, allowances can be sold or traded nationwide.

The allowance program works better than previous programs because it is a reward system. Power companies with emissions below the Clean Air Act's allowable level can sell their extra permits for profit. This gives companies the incentive to conserve energy and make technological innovations, as well as to encourage their customers to conserve energy. Investing in these incentives means the company becomes more profitable. Companies that install new pollution controlling devices are also given bonus allowances. Improvements include installing devices that use clean coal technology, since many power plants in the nation still burn coal to produce energy.

Cleaner Coal

There are several ways in which coal can be made to burn "cleaner." Coal has a sulfur content between 1 and 6 percent, and when it is burned, that sulfur is released as sulfur dioxide. In "clean" coal some or most of the sulfur has been removed. Coal can be cleaned by physical, chemical, and biological means.

Coal is physically cleaned by crushing it and removing the impurities. The coal is crushed until it has the consistency of sand. Then it is sprayed with jets of water and sifted through several screens. The bits of rock, dirt, and other impurities are removed by screens. Last, the small pieces of coal are spun rapidly in a machine, separating the remaining impurities and pushing them to the bottom of the machine. Twenty to fifty percent of the sulfur content can be removed in this manner.

Chemicals and bacteria can also be used to remove the sulfur from the coal. Adding chemicals or a particular species of bacteria to the coal changes its chemical composition. The sulfur is released from the coal and is easier to remove. Both these processes remove 10 to 30 percent of the sulfur in the coal.

Finally, coal can be cleaned while it is being used to produce electricity. With the proper machinery, limestone can be mixed with the coal while it is being burned. The limestone absorbs sulfur and falls to the bottom of the incinerator as ash. Adding oxygen to the burning coal can also help by turning the burnt coal to liquid or gas and leaving behind a sulfur residue that can be cleaned up afterward. These effective but expensive methods decrease the sulfur dioxide released by 90 to 95 percent.

A worker walks among mounds of clean coal at an Ohio processing plant. Cleaning coal removes some or most of its sulfur, making it safer to burn for fuel.

Most of these procedures are still new technology. The more common method of cleaning coal is to use a "scrubber" in the plant's smokestacks. The scrubber is a smoke-catching device installed on the smokestack. In the scrubber the gases released are sprayed with water and lime. The sulfur dioxide mixes with the lime and falls to the bottom of the scrubber as removable sludge.

This process is expensive, but effective, removing almost 95 percent of the sulfur dioxide emitted by burning coal. Scrubbers can cost $300 million dollars apiece and cost a great deal to maintain and operate. There is also the problem of the leftover sludge; it is classified as hazardous waste and must be disposed of properly. Today all of the newer power plants include scrubbers, but scientists continue to research less expensive means to clean coal that are still effective.

Cleaner Transportation

In the United States, motor vehicles are responsible for more than 50 percent of nitrous oxide pollution. Because of these statistics, part of the Clean Air Act focuses on the problem of pollution from vehicular use. The Clean Air Act addresses cleaner fuel, more efficient cars, and the reduction of vehicular use.

The biggest problem today with vehicular use is the number of vehicles on the road. According to the Environmental Protection Agency, "In 1970, Americans traveled one trillion miles in motor vehicles and are expected to travel four trillion miles each year by 2000."[38] Most people also still drive alone, even when car pools and other alternatives are available. To combat this, in some metropolitan areas transportation policies have been enacted to discourage single occupancy in vehicles. For example, states are allowed to add surcharges to parking fees for single occupancy. If it is expensive to park, the reasoning goes, people may choose to share a ride.

The 1990 Clean Air Act contains regulations to create cleaner fuels. Gasolines now must contain detergents that

Designing cars like this Toyota that are powered by electric or partially electric engines helps to reduce the amount of nitrous oxide pollution in the air.

prevent the buildup of deposits in engines. Engine buildup causes fuel to burn less cleanly, releasing more emissions. The regulations also encourage development and sale of alternative fuels such as alcohols, liquefied petroleum gas, and natural gas.

Cars must now run cleaner than in the past as well. Cars have dashboard warning lights to alert drivers if the pollution control devices stop working. In the past these devices had to be built to last 50,000 miles, but today they must be built to last 100,000 miles. Manufactures must now include a certain number of clean fuel cars—cars with electric or hybrid engines—in their overall production. In California at least five hundred thousand clean fuel cars have to be manufactured for sale each year.

By lowering the number of nitrous oxides produced by vehicles and the amount of sulfur dioxide from power plants, the Clean Air Act hopes to reduce polluting emissions 50 percent by 2010. When the amendments were passed in 1990, many politicians and scientists felt that these were impossible goals and that they would be very expensive to achieve. To ensure that the government was working in the right direction, the Clean Air Act allocated funds to measure acid pollution.

National Acid Deposition Program

The National Acid Deposition Program (NADP) is a network of over two hundred precipitation monitoring sites. These sites span the continental United States, Alaska, Puerto Rico, and the Virgin Islands. All of these sites work together, reporting data that can be used for research and to help policy makers make informed decisions.

All of the sites of the NADP work exactly the same. Every week samples are taken from the collection devices at the sites. Both air and rain samples are collected and measured. Strict "clean-handling" procedures are enforced, meaning researchers take extra care to be certain their data is not tainted. The samples are then sent to a central laboratory where they are analyzed for sulfate, nitrate, and other pollutants.

By accessing the NADP's website, anyone can see the results of these tests. In this way everyone can see how the Clean Air Act is working to change emissions in the environment.

Success

Although the year 2010 has not yet arrived, the results of the Clean Air Act Amendments so far are impressive. The Environmental Protection Agency used the latest NADP emissions data and a sophisticated array of computer models in 1990 to project what the year 2010 will bring. According to a press release from the EPA, "in the year 2010 the amendments of 1999 will prevent 23,000 Americans from dying prematurely, and avert over 1,700,000 incidences of asthma attacks and aggravation of chronic asthma."[39] The EPA also projects that many other pollution-related hospitalizations will be prevented. The EPA's speculation of the amount of deaths and illnesses that will be prevented by the 1990 Clean Air Act Amendments also includes a cost analysis. Many opponents of the regulations felt that the cost of the program would far outweigh the benefits. The EPA projects that the amount of money saved from health and ecological benefits of the program is approximately $110 billion. In comparison, the cost of the Clean Air Act is $27 billion. Therefore, the proponents of the measures contend that benefits outweigh the costs by four to one.

Many environmentalists point out that there is still much work to be done. The success of the Clean Air Act is only projected. Still, there is no arguing that sulfur dioxide emissions have decreased over the last decade. The lowered emissions are helping, but in some areas the damage has already been done. Many researchers are studying ways to revive acid-damaged lakes and forests.

Undoing the Damage

Lakes and forests that are acidified will not rejuvenate quickly despite recovery measures. Once high concentrations of acids are a part of the ecosystem, they take a very long time to diminish. Scientists speculate that the chemical recovery of an ecosystem could take decades.

In order to speed up the process of chemical recovery, some countries, including the United States, have resorted to liming water bodies and soils. Liming is the process of adding ground-up limestone to waters or soils. The lime in the limestone is a calcium-based material that neutralizes acids. When lime is added, acid levels decline and the ecosystem may return to a healthier pH.

Sweden, Norway, and Canada have been liming since the 1970s, and although they have had positive results, they have also discovered that liming has drawbacks. The primary problem is the expense. The amount of lime required for acid neutralization is immense and therefore expensive. It is estimated it would cost $750,000 just to lime all of the acidified lakes in the Adirondack Mountains and the northeast United States.

Liming also does not address the problem of heavy metals in the environment. Levels of acidity may return to normal, but the metals released by acids must diminish before plants and animals will be healthy enough to survive.

Swedish authorities pump tons of lime into an acidified lake. Liming is an effective, but expensive, way to reduce a lake's pH level.

Scientists are uncertain how long this biological recovery will take. Since there have been few long-term studies that monitor recovery of acidified ecosystems, they can only speculate about the length of time. Researcher Charles Driscoll is unsure how much time a forest would require to recover, but, he states, "it is likely to be at least decades after soil chemistry is restored because of the long life of tree species and the complex chemical reactions of soil, roots, microbes and soil biota."[40] He also believes that it would take at least ten years for zooplankton to recover in a lake and then another five to ten years after that before the fish would return.

Many researchers are also concerned about the sensitivity of these previously damaged ecosystems. Although pollution has declined dramatically, emissions are still much higher than would be naturally occurring in the environment. These previously acidified ecosystems are unable to buffer acids. This means that newer, lower levels of acid in the rain can harm them again.

A Need for Tougher Regulations

Environmentalists do not argue that the Clean Air Act emissions reductions have made a positive impact, but

A sign in Nova Scotia warns of the danger acid rain poses to rivers. Despite progress in the fight against acid rain, environmentalists contend that more needs to be done to fully eradicate the problem.

they do argue that the changes are not enough. According to the Clean Air Task Force, even with the current regulations, "nearly 10,000 Canadian lakes will be damaged . . . thirty percent of brook trout streams in Virginia will not be able to support brook trout . . . [and] declining vigor of red spruce and sugar maple will likely occur in other tree species as well."[41] The Science Links Project of the Hubbard Brook Research Foundation concludes that in order for ecosystems to begin recovering fully, emissions must be cut a further 80 percent from the level established by the 1990 Clean Air Act Amendment.

Since Canada receives much of its pollution from the United States, the Canadian government also feels that further reduction of emissions is crucial to the recovery of their damaged ecosystems. The Canadian Acidifying Emissions Task Group has recommended a 75 percent further reduction from the United States. Without this reduction the researchers believe that seventy-six thousand lakes in southeastern Canada will remain damaged.

These environmental groups argue that the solution to high emissions can be solved through affordable technology. Now that cleaner coals are available, more power plants should be using them. Environmentalists also argue that scrubbing coals is highly effective and all power plants should use this technology. Although most power plants now use scrubbers, some older power plants do not. If this changes, emissions will be even lower.

Global Pollution

Much of the worldwide efforts to combat acid pollution are similar to the tactics used in the United States. European countries have joined together to lower emissions throughout the continent. Europeans have formed the UN Economic Commission for Europe that has drafted an agreement to lessen pollution. Their regulations on the amount of sulfur dioxide and nitrous oxides emitted are similar to those of the United States.

It is crucial that the countries work together since they share pollution across the borders. The same is true of the

United States and Canada, where transboundary pollution continues to be a problem. Scientists and politicians both agree that pollution is a worldwide problem that must be addressed by all countries that produce and share sulfur and nitrogen pollution.

The countries that produce pollution have been traditionally thought of as the industrial countries of Europe and North America, but in recent years more countries have been added to the list of polluters. In South America and Asia there are many areas that have recently become industrialized. There are no regulations in these new areas of industry, and these countries are resistant to instituting emissions laws. Pollution regulations would decrease their profitability. These countries are growing so rapidly that scientists are worried these countries may ultimately contribute a tremendous amount of pollution to the planet.

Asia's growth is so rapid that it is believed the energy needs of the continent will greatly deteriorate the air quality. It is estimated that by the year 2010 there will be over 4 billion people living in eastern Asia and the Indian subcontinent. More people means there is a greater demand for energy, and the primary source for energy is coal. These countries do not yet have the emission control devices of Europe and North America, and the sulfur dioxide emissions there are quickly increasing.

The acid rain problem in Asia has already begun. The Chinese Research Academy of Environmental Sciences reports that "40% of China is affected by acid rain causing $1.6 billion dollars worth of damage to crops, forests and property annually."[42] Scientists agree that unless the countries of Asia and other newly industrialized countries learn from the mistakes of Europe and North America, they may find themselves in a similar, if not worse, situation.

The Future

Most environmentalists agree that the United States is on the right track to recovery. However, most scientists also agree that more research must be done before acid rain becomes a problem of the past. At a 2001 acid rain conference, Congressman Sherwood Boehlert stated, "The response to acid rain has thus far been a story of the system succeeding—a story of Congress acting on the basis of the best research, even in the face of uncertainty, and then trying to update policy on the basis of continuing research. That's a pattern we need to perpetuate. We need to keep funding the research and heeding its results."[43]

Many scientists would agree with Boehlert's conclusion. They argue that the research over the last decade shows that the various laws still have not done enough to decrease the damage that acid rain is causing. Brian McLean of the U.S. Environmental Protection Agency stated, "I look at what we know today that we didn't know ten years ago when we last acted on this issue. I look at the progress we have made in both science and in our ability to manage multi-state air pollution problems. We have shown that benefits of additional reduction of sulfur dioxide and nitrous oxides far exceed additional costs."[44]

Environmentalists also point to the fact that humans rely too heavily on fossil fuels. Fossil fuels are a limited resource, which will someday no longer be available. The burning of fossil fuels is also the largest contributor to pollution. Environmentalists argue that alternate sources of energy need to be researched and developed for the future.

Perhaps the most important part of the solution to acid rain and other forms of acid pollution is the education of the public. The more knowledgeable people are about the causes and problems of pollution, the easier it is for them to be conscious of environmental issues and to work toward a solution. If children learn about the issues surrounding pollution in school, they may grow up to be a part of the solution themselves. The more people understand about acid rain, the more likely they are to contribute to its end, and that may be what ultimately makes the difference.

Notes

Chapter One: The Acid Rain Problem

1. Quoted from The Pennsylvania State Archives. www.docheritage.state.pa.

2. John McCormick, *Acid Earth: The Politics of Acid Pollution.* London: Earthscan, 1997, p. 3.

3. McCormick, *Acid Earth*, p. xv.

4. Quoted in Hillary Mayell, "Scientists Use Tombstones to Track Environmental Changes," *National Geographic News*, November 27, 2001. www.news.nationalgeographic.com.

5. McCormick, *Acid Earth*, p. 36.

6. Chris LeRose, "The Collapse of the Silver Bridge," *West Virginia Historical Society Quarterly*, vol. xv, no. 4, October 2001. www.wvculture.org.

7. Quoted in Jurgen Schmandt and Hilliard Roderick. *Acid Rain and Friendly Neighbors: The Policy Dispute Between Canada and the United States.* Durham, NC: Duke University Press, 1985, p. 47.

8. Ellen Baum, *Unfinished Business: Why the Acid Rain Problem Is Not Solved.* Clean Air Task Force, October 2001, p. 6. www.cleartheair.org.

9. McCormick, *Acid Earth*, p. 32.

10. Brian A. Forster, *The Acid Rain Debate: Science and Special Interests in Policy Formation.* Ames: Iowa State University Press, 1993, p. 21.

Chapter 2: Dying Lakes

11. Tom Frost, "UW-Madison Project Shows Persistence of Acid Rain Effects," University of Wisconsin-Madison, August 2000. www.news.wisc.edu.

12. Baum, *Unfinished Business*, p. 4.

13. Swedish Environmental Protection Agency, "Facts About Swedish Policy: Acid Rain," September 1999, p. 2. www.environ.se.

14. Forster, *The Acid Rain Debate*, p. 63.

15. W. Swenson and T. May, *Wisconsin Fisheries and Acid Rain*, Department of Agricultural Journalism, University of Wisconsin-Extension, 1987, p. 3.

16. W. Swenson and T. May, *Wisconsin Fisheries and Acid Rain*, p. 4.

17. Baum, *Unfinished Business*, p. 4.

18. McCormick, *Acid Earth*, p. 30.

Chapter 3: Dying Forests

19. Quoted in *PR Newswire*, "Fall Foliage and Acid Rain in Pennsylvania: Where Have All the Colors Gone? Asks the Clean 'Em Up Campaign," October 7, 1999. www.findarticles.com.

20. Quoted in *PR Newswire*, "Fall Foliage."

21. Chris Rose, "The Forest Is Dying," *New Internationalist*, June 1988. www.newint.org.

22. McCormick, *Acid Earth*, p. 22.

23. Baum, *Unfinished Business*, p. 7.

24. Robert Winkler, "Is Acid Rain Killing Off Wood Thrushes?" *National Geographic News*, August 13, 2002. http://news.nationalgeographic.com.

Chapter 4: How Acid Rain Affects Humans

25. Quoted in John Neilsen, "The Killer Fog of '52," *National Public Radio*, December 11, 2002. www.npr.org.

26. Quoted in Neilsen, "The Killer Fog."

27. Minnesota Pollution Control Agency, "Appendix C: Criteria Pollutants," February 16, 2001. www.pca.state.mn.us.

28. Quoted in April Reese, "Blue Skies," *E Magazine*, November/December 1999. www.emagazine.com.

29. Senator Cheryl A. Jacques, "Attackingsthma: Combating an Epidemic Among Our Children," Massachusetts Senate Committee on Post Audit and Oversight, December 2002. www.state.ma.us.

30. Quoted in Jacques, "Attacking Asthma."

31. T. May and L. Klessig, *Acid Precipitation's Impact on Materials, Visibility and Human Health*, Department of Agricultural Journalism, University of Wisconsin-Madison, 1990, p. 7.

32. Health Canada, "Aluminum and Human Health," Canada Health Water Talk, April 2003. www.hc-sc.gc.ca/hecssescs/water.

33. Sharon L. Crenson, "Research of Mercury Contamination Leaves Huge Gap in Knowledge," Great Lakes Environmental Directory, October 9, 2002. www.greatlakesdirectory.org.

34. Quoted in Crenson, "Research of Mercury."

Chapter Five: Searching for Solutions

35. McCormick, *Acid Earth*, p. 143.

36. McCormick, *Acid Earth*, p. 142.

37. United States Congress, "Clean Air Act," Environmental Protection Agency, website, 1990. www.epa.gov.

38. United States Environmental Protection Agency, "Mobile Sources," *The Plain English Guide to the Clean Air Act*, 1990. www.epa.gov.

39. Dave Ryan, "New Report Shows Benefits of 1990 Clean Air Act Amendments Outweigh Costs by Four to One Margin," United States Environmental Protection Agency, November 16, 1999. www.epa.gov.

40. Charles Driscoll et al., "Acidic Deposition in the Northeastern United States: Sources and Inputs, Ecosystem Effects and Management Strategies," *Bioscience*, vol. 51, no. 3, 2001, p. 194.

41. Baum, *Unfinished Business*, p. 1.

42. Quoted in Gregory R. Carmichael and Richard Arnott, "Baseline Assessment of Acid Deposition in Northeast Asia," Nautilus Institute, 1996. www.nautilus.org.

43. Quoted in James C. White, "Acid Rain: Are the Problems Solved?" Washington, DC: Center for Environmental Information, Inc., May 2–3, 2001, p. x. www.rochesterenvironment.org.

44. Quoted in White, "Acid Rain," p. iv.

Glossary

acidification: The process of making lakes, forests, or other ecosystems excessively acidic through acid deposition.

acid deposition: When acids fall from the air and deposit on the ground, buildings, or living things.

acid neutralizing capacity: The ability of water or soil to naturally neutralize acids.

aluminum: A silvery-white element that is commonly found in the earth's crust. When released in soil and water, it is believed to be toxic.

Alzheimer's disease: A disease that affects the brain, first with impaired memory, followed by impaired thought and speech, and finally complete helplessness.

anemia: A lack of red blood cells in the body that causes weakness and fatigue.

asthma: A chronic respiratory disease, often arising from allergies, that is characterized by sudden recurring attacks of labored breathing, chest constriction, and coughing.

bioaccumulation: The accumulation of a substance, such as a toxic chemical, in various tissues of a living organism.

calcium: An element that is commonly found in the earth's crust that contributes to growth, bone development, and many other processes in living organisms.

Clean Air Act: A series of regulations in the United States that have the purpose of diminishing pollution.

conifers: Any of various needle or scale-leaved trees such as evergreens, cone-bearing trees, or shrubs such as pines, spruces, and firs.

dead lake: A lake that can no longer sustain life due to acidification.

deciduous: A plant having foliage that is shed annually at the end of the growing season.

dry deposition: Acids that fall to the earth in tiny dry particles.

ecosystem: An ecological community functioning as a unit within a particular environment.

emissions allowance: A system used by the Clean Air Act that gives power plants a certain number of allowances for pollution emitted.

fossil fuels: Fuels such as petroleum, coal, or natural gas that are derived from the remnants of ancient living matter buried in the earth.

gills: The organ used by fish to breath.

heavy metals: An elemental metal that is usually poisonous.

liming: The process of grinding up limestone and adding it to water or soil in order to lower acidity.

mercury: A silver-white, poisonous, metallic element.

National Acid Deposition Program: A U.S. program that measures the amount of acid deposition at over two hundred research stations across the nation and its island holdings.

pH scale: The scale used by scientists to determine levels of acidity.

smog: A combination of the words "fog" and "smoke," used to describe polluted air.

superstack: Tall smoke stacks used by factories and power plants to put smoke high into the air and away from the people on the ground.

Waldsterben: A German word literally meaning "forest death," used to describe the phenomenon of dying forests of Europe.

wet deposition: Acid falling to the earth in the form of precipitation such as rain, snow, sleet, or fog.

Organizations to Contact

The Adirondack Council
342 Hamilton Street, Albany, NY 12210
800-842-PARK
info@adirondackcouncil.org
www.adirondack.org

The Adirondack Council is a nonprofit environmental group that has been working since 1975 to protect the open-space resources of New York State's 6 million-acre Adirondack Park. The council also seeks to help sustain the natural and human communities of the region.

American Lung Association
61 Broadway, 6th Floor, NY, NY 10006
212-315-8700
www.lungusa.org

The American Lung Association works toward the cure and prevention of lung diseases. The association is involved in trying to help solve the problem of asthma and to end acid pollution.

Environment Canada
351 St. Joseph Boulevard, Hull, Quebec K1A 0H3
819-997-2800 or 1-800-668-6767
enviroinfo@ec.gc.ca
www.ec.gc.ca

Environment Canada works to preserve the environment of Canada including water, air, and soil quality. Its mission is to help Canadians live and prosper in an environment that needs to be respected, protected, and conserved. Environment Canada reports and responds to problems of acid rain.

National Atmospheric Deposition Program
Illinois State Water Survey, 2204 Griffith Drive,
Champaign, IL 61820-7495
217-333-7873

http://nadpsws.uiuc.edu

There are over two hundred NADP test sites in the United States that take measurements on the acidity of water and air. By accessing the NADP website, anyone can view the results of their research as well as learn more about acid rain.

United States Environmental Protection Agency Acid Rain Program
Clean Air Markets Division, 1200 Pennsylvania Avenue, NW, Mail Code 6204N, Washington, DC 20460
202-564-9150

www.epa.gov

The overall goal of the Acid Rain Program is to achieve significant environmental and public health benefits through reductions in emissions of sulfur dioxide and nitrogen oxides, the primary causes of acid rain.

USDA Forest Service
P.O. Box 96090, Washington, D.C. 20090-6090
(202) 205-8333

www.fs.fed.us

Established in 1905, the Forest Service is an agency of the U.S. Department of Agriculture. The Forest Service manages public lands in national forests and grasslands.

Suggestions for Further Reading

Books

Alex Edmonds, *Acid Rain.* Brookfield: Cooper Beech Books, 1997. This concisely written book examines the causes and effects of acid rain and discusses its prevention. It is full of photographs.

Tony Hare, *Acid Rain.* London UK: Glouchester Press, 1990. This is a well-illustrated book that covers the basic issues surrounding acid rain.

Eileen Lucas, *Acid Rain.* Chicago: Childrens Press, 1991. Covering the history of acid rain and focusing on the United States, this book is a good overview of the acid rain issue.

Sally Morgan, *Acid Rain.* New York: Franklin Watts, 1999. A brief overview of the problems that acid rain causes, with good photography of recent incidents and issues.

Cass R. Sandik, *A Reference Guide to Clean Air.* Hillside, NJ: Enslow, 1990. This book contains an alphabetical listing of nearly two hundred names, terms, and concepts that are related to air pollution, with simple explanations of their meanings and relevance.

Peter Tyson, *Acid Rain.* New York: Chelsea House, 1992. This book has in-depth coverage of the problems and potential solutions of acid rain.

Websites

Kids Connect (http://www.kidskonnect.com). Kidskonnect is an educational website for children and young adults designed by educators. This site has a good section on acid rain that can be used for reports or just to learn more about the issue.

Environment Canada's Acid Rain Site (http://www.ec.gc.ca/acidrain). Environment Canada is an organization run by the Canadian government that is responsible for environmental legislation. This site has a lot of great information on acid rain and includes a page as well.

National Atmospheric Deposition Program (http://nadp sws.uiuc.edu). This website reports the current results of the NADP research. Information on acidity throughout the United States can be accessed from this site.

United States Environmental Protection Agency Acid Rain Homepage (www.epa.gov). This website explains current U.S. government policies regarding acid rain and its reduction.

Works Consulted

Books

Bruce A. Forster, *The Acid Rain Debate:Science and Special Interests in Policy Formation.* Ames: Iowa State Press, 1993. Professor Forster looks at both the social and economic issues surrounding the environmental issue of acid rain.

Russell W. Johnson et al., *The Chemistry of Acid Rain.* Washington DC: American Chemical Society, 1987. A collection of scientific papers from North American scientists studying how acid rain is created.

Allan H. Legge and Sagar V. Krupa, *Acidic Deposition: Sulphur and Nitrogen Oxides.* Chelsea, MI: Lewis, 1990. A collection of scientific papers that research the effects of acid rain in Canada.

———, *Acid Rain: Its Causes and Its Effects on Inland Waters.* Oxford, UK: Clarendon Press, 1992. A book that examines the effects of acid rain on lakes and streams and the wildlife that lives in acidified waters.

B.J. Mason, *The Surface Waters Acidification Programme.* Cambridge, UK: Cambridge University Press, 1990. A collection of the studies done in Europe by the Surface Waters Acidification Programme.

John McCormick, *Acid Earth: The Politics of Acid Pollution.* London: Earthscan, 1997. This book explains in simple terms the science and politics that surround the issue of acid rain.

Laurence J. O'Toole Jr., *Institutions, Policy and Outputs for Acidification: The Case of Hungary.* Aldershot, UK: Ashgate, 1998. A book that focuses on Hungary's example of implementing international regulations regarding pollution and acid rain.

Jurgen Schmandt and Hilliard Roderick, *Acid Rain and Friendly Neighbors: The Policy Dispute Between Canada and the United States.* Durham, NC: Duke University Press, 1985. An examination of the joint effort of Canada and the United States to solve the problem of acid rain.

William E. Sharpe and Joy R. Drohan, *The Effects of Acid Deposition on Pennsylvania's Forests.* University Park, PA: Environmental Resources Research Institute, 1999. This book contains the papers presented at the 1998 Pennsylvania Acidic Deposition Conference, representing the most recent research on how acid rain harms forests.

Timothy J. Sullivan, *Aquatic Effects of Acid Deposition.* Boca Raton, FL: Lewis, 2000. A summary of the advancements in research since 1990 regarding the effects of acid rain on lakes and streams.

Periodicals
"Japan Cooperates with China to Prevent Acid Rain," *Asian Economic News*, October 23, 2000.

Paul Beck, "Cleaning Up King Coal," *Popular Science*, June 19, 2002.

Nancy Carson, "Flaws in Conventional Wisdom on Acid Deposition. (National Acid Precipitation Assessment Report)," *Environment*, March 2000.

Charles Driscoll et al., "Acidic Deposition in the Northeastern United States: Sources and Inputs, Ecosystem Effects and Management Strategies," *Bioscience*, vol. 51, no. 3, 2001.

Malcolm Fergusson, "Greening Transportation," *Environment*, January–February 1999.

Sarah Graham, "Acid Rain Linked to Bird Decline," *Scientific American*, August 13, 2002.

Kristin Leutwyler, "Nitric Oxide Proves a Potent Pollutant," *Scientific American*, December 6, 2000.

Don Munton, "Dispelling the Myths of the Acid Rain Story," *Environment*, July–August 1998.

Richard V. Pouyat, "Science and Environment Policy: Making Them Compatible," *Bioscience*, vol. 49, no. 4, 1999.

April Reese, "Bad Air Days (Air Pollution in the United States)," *E Magazine*, November 1999.

Marvin Soroos, "Preserving the Atmosphere as a Global Commons," *Environment*, March 1998.

Jorgen Wettestad, "Clearing the Air: Europe Tackles Transboundary Pollution," *Environment*, March 2002.

Reports
Ellen Baum, *Unfinished Business: Why the Acid Rain Problem Is Not Solved*, Clean Air Task Force, October 2001. www.cleartheair.com

C.T. Driscoll et al., *Acid Rain Revisited: Advances in Scientific Understanding Since the Passage of the 1970 and 1990 Clean Air Act Amendments*, Hubbard Brook Research Foundation, 2001.

R. Fischer et al., *The Condition of Forests in Europe: 2002 Executive Report*, Federal Research Centre for Forestry and Forest Products, 2002.

T. May and J. Eilers, *Acid Rain: Impact on Aquatic Organisms Other Than Fish*, Department of Agricultural Journalism, University of Wisconsin-Madison, 1987.

T. May and L. Klessig, *Acid Precipitation's Impact on Materials, Visibility and Human Health,* Department of Agricultural Journalism, University of Wisconsin-Madison, 1990.

Hans Martin Seip, *Acid Rain and Climate Change; Do Environmental Problems Have Anything in Common?* Center for International Climate Control and Research, June 2001.

William E. Sharpe et al., *The Effects of Acidic Deposition on Aquatic Ecosystems in Pennsylvania*, Environmental Resources Research Institute, 1999.

W. Swenson and T. May, *Wisconsin Fisheries and Acid Rain*, Department of Agricultural Journalism, University of Wisconsin-Madison, 1987.

Internet

Phil Brown and Corydon Ireland, "Acid Rain Won't Go Away," *Adirondack Explorer*, June 2001. www.adirondackexplorer.com.

Canadian Association of Physicians for the Environment, "Implications for Human Health: Acid Deposition and Transregional Pollution," April 1995. www.cape.ca.

Gregory R. Carmichael and Richard Arnott, "Baseline Assessment of Acid Deposition in Northeast Asia," Nautilus Institute, 1996. www.nautilus.org.

CNN Interactive, "Deadly Smog 50 Year Ago in Donora Spurred Clean Air Movement," Department of Environmental Protection Pennsylvania, October 27, 1998. www.dep.state.pa.us.

Sharon L. Crenson, "Research of Mercury Contamination Leaves Huge Gap in Knowledge," Great Lakes Environmental Directory, October 9, 2002. www.greatlakesdirectory.org.

Doc Heritage, "The Donora Smog Disaster October 30–31, 1948," Pennsylvania State Archives. www.docheritage.state.pa.us.

Ecological Society of America, "Acid Deposition," 2000. http://esa.sdsc.edu.

Environment News Service, "Acid Rain Leads to Songbird Declines," August 16, 2002. http://ens-news.com.

Tom Frost, "UW-Madison Project Shows Persistence of Acid Rain Effects," University of Wisconsin-Madison, August 2000. www.news.wisc.edu.

Health Canada, "Aluminum and Human Health," Canada Health Water Talk, April 2003. www.hc-sc.gc.ca/hecs sescs/water.

Jay Henderson, "Crisis Looming in Virginia's Trout Streams," West Virginia Highlands, 1998. www.wvhighlands.org.

Chris LeRose, "The Collapse of the Silver Bridge," *West Virginia Historical Society Quarterly*, vol. xv, no. 4, October 2001. www.wvculture.org.

Hillary Mayell, "The Environmental Movement at 40: Is the Earth Healthier?" *National Geographic News*, April 19, 2002. http://news.nationalgeographic.com.

Elaine McGee, "Acid Rain and Our Nation's Capital," USGS, 1997. http://pubs.usgs.gov.

Minnesota Pollution Control Agency, "Appendix C: Criteria Pollutants," February 16, 2001. www.pca.state.mn.us.

National Geographic Today, "Pollution Over South Asia Threatens Economies," *National Geographic News*, August 12, 2002. http://news.nationalgeographic.com.

John Neilsen, "The Killer Fog of '52," *National Public Radio*, December 11, 2002. www.npr.org.

Norwegian Pollution Control Authority, "State of Environment Norway: Acid Rain," January 2003. www.environment.no.

PR Newswire, "Fall Foliage and Acid Rain in Pennsylvania: Where Have All the Colors Gone? Asks the Clean 'Em Up Campaign," October 7, 1999. www.findarticles.com.

April Reese, "Blue Skies," *E Magazine*, November/December 1999. www.emagazine.com.

Chris Rose, "The Forest Is Dying," *New Internationalist*, June 1988. www.newint.org.

Dave Ryan, "New Report Shows Benefits of 1990 Clean Air Act Amendments Outweigh Costs by Four to One Margin,"

United States Environmental Protection Agency, November 16, 1999. www.epa.gov.

Swedish Environmental Protection Agency, "Facts About Swedish Policy: Acid Rain," September 1999. www.environ.se.

United States Congress, "Clean Air Act," Environmental Protection Agency Website, 1990. www.epa.gov.

United States Environmental Protection Agency, Mobile Sources, *The Plain English Guide to the Clean Air Act*, 1990. www.epa.gov.

James C. White, "Acid Rain: Are the Problems Solved?" Washington, DC: Center for Environmental Information, Inc., May 2–3, 2001. www.rochesterenvironment.org.

Robert Winkler, "Is Acid Rain Killing Off Wood Thrushes?" *National Geographic News*, August 13, 2002. http://news.nationalgeographic.com.

Index

acid deposition, 13
acid neutralizing capacity, 25
acid pollution and rain
  continuing importance of, 8–9
  creation of, 13–14
  discovery of, 6
  Donora fog and, 10–12
  history of, 14–15
  origin of, 6
  traveling across local and national boundaries and, 21–22
  see also, damage from acid rain; solutions, to acid rain
Adirondack Mountains (New York), 42
algae, 32–33
aluminum
  fish population and, 32
  forest soil and, 43–44
  health risks and, 55–56
Alzheimer's disease, 55
American Steel and Wire Company, 10
amphibians, 28
anemia, 56
animals
  acid rain's indirect effect on, 33–34
  forest, 45–46
  metals in, 54–55
Appalachian forests, 44
aquatic environments, 19–20, 28–29
aquatic plants, 32–33
Asia, 72
asthma, 50–53
automobiles, 63, 66–67
autumn foliage, 35–36

Bainbridge, Lois, 12
bioaccumulation, 54
birds
  aquatic, 33–34
  in forest ecosystem, 45
Blue Ridge Mountains, 41
Bush, George H.W., 62–63

calcium, in forest soil, 43
Canada
  acid pollution from U.S. to, 22
  liming in, 69
  sugar maple forests in, 43
children
  asthma and, 51–53
  mercury poisoning and, 57
China, 72
Clean Air Act (1970), 20
Clean Air Act Amendments, 8, 62–63, 66–67, 68

coal
  cleaning of, 64–65
  London Killer Fog and, 48
  sulfur dioxide and, 13
companies
  pollution allowances for, 63–64
  tall smokestacks built by, 20–21
conifer trees, 41–42
copper piping, 59
Cribb, Stan, 47

damage, from acid rain
  in drinking water, 58–59
  to human health, 55–56
  indirect effect of, on animals, 33–34
  lake studies on, 23–25
  lung, 49–51
  mercury poisoning as, 56–58
  to metals, 16–18
  to plants and animals, 18–20
  to stones and monuments, 16
  see also forests; lakes
deaths
  from asthma, 53
  Donora fog and, 11
  from killer smog, 20
  London Killer Fog and, 48, 49
  from Ohio River Silver Bridge collapse, 18
deciduous trees, 38–39
DeRidder, Kim J., 19
dipper, 33–34
diseases, 55–56
Donora, Pennsylvania, 10–12
drinking water, 58–59

earthworms, 45
ecosystems
  chemical recovery of, 68–69
  forest, 44–45
  lake, 27–28
Environmental Protection Agency (EPA), 68
Europe
  combating acid pollution in, 71
  lakes in, 26–27
  see also names of individual countries

Faroe Islands, 57
fingerlings, 29–30
fish
  acidified water's effect on, 31–32
  disappearance of lake, 29
  mercury poisoning and, 57
  metals in, 54
fog
  acidification in forests and, 41
  breathing acid into lungs and, 49–50
  Donora Fog (1948), 10–12
  see also smog
food, contaminated, 53–55
Food and Drug Administration (FDA), 57

forests
  acid fog and, 41
  coating on tree leaves of, 41
  conifer needles in, 41–42
  destruction of European, 38–40
  killing of ecosystems in, 44–45
  Pennsylvania foliage and, 35–36
  poisoning of soil in, 43–44
  survival of tree seedlings in, 42–43
  susceptibility to acidification in, 40–41
  undoing damage to, 68–70
fossil fuels, 12–13, 73
frog eggs, 28

gasoline, 66–67
Germany, 37–39
granite bedrock, 26–27
gravestones, 16
Great Britain, 22
Green Mountains (Vermont), 42

Harvey, Harold, 29, 30
health risks, 55–56

International Nickel Company (INCO), 21
inversion, 10
lakes
  acidity effecting ecosystems in, 27–28
  aquatic life in, 28–29
  fish population in, 31–32
  plant life in, 32–33
  resistance of, to acids, 25–26
  study on effects of acid rain on, 23–25
  undoing damage to, 68–70
  as vulnerable to acidification, 26–27
legislation
  need for tougher, 70–71
  under Reagan, 60–62
  see also Clean Air Act
liming process, 69
Little Rock Lake (Wisconsin), 23–25
London Killer Fog (1952), 47–49
loons, 33
Lou Gehrig's disease, 55–56
Lumsden Lake experiment, 29–30

mercury poisoning, 56–58
metals
  damage of acid rain to, 16–18
  in food chain, 53–55
  see also aluminum
monuments, 16
Mount, Mitchell, 41

National Acid Deposition Program, 67–68
nitrogen oxides, 13, 33, 49–50
Norway, 33–34, 69

Oden, Svante, 22
Ohio River Silver Bridge, 17–18

Paine Run River, 19–20
Parkinson's disease, 55–56
Parthenon (Greece), 16
Pennsylvania State University, 35–36
pH levels
  fish population and, 31
  forest soil and, 44, 45
  in lakes, 30
  Little Rock Lake study and, 24
  scale measuring of, 15
plankton, 28
politicians, 7
power plants, pollution allowances for, 63–64
pregnant women, 57–58

Reagan, Ronald, 60–62
red spruce, 42
regulation. See Clean Air Act; legislation
research
  on mercury poisoning, 57–58
  need for further, 72–73
  on Pennsylvania forests, 35–36
  under Reagan, 60, 61
respiratory illnesses, 50–51
rivers, damage to, 19–20

salmon, 29–30
Scandinavia, 26–27

scientists, 7
silver fir trees, 37–39
Smith, Robert Angus, 14–15
smog
  deaths from "killer," 20
  London Killer Fog and, 47–49
  see also fog
smokestacks, 20–22
smokestack scrubbers, 65
solutions, to acid rain
  agreeing on course of action for, 7–8
  as a global task, 71–72
  liming water bodies and soils as, 69
  precipitation monitoring sites as, 67–68
  public awareness of, 73
  reducing car pollution as, 66–67
  reducing emissions from power plants as, 63–64
  tall smokestacks and, 20–22
  U.S.–Canadian disagreements on, 60–62
  see also Clean Air Act
South America, 72
sphagnum moss, 24
stone, 16
studies. *See* research
sugar maple tree, 36, 43
sulfur dioxide, 13
  auto emissions and, 63
  cleaning coal and, 65
  lung damage and, 49–51

Susquehannock State Forest, 36
Sweden, 22, 69

Taj Mahal (India), 16
tulip poplar, 44

University of Wisconsin–Madison, 23

"Waldsterben," 38–40
weather, 40–41
wet acid deposition, 13–14
whales, 58
white suckers, 30
wood thrushes, 45

zinc mill, 10–12
zooplankton, 32

Picture Credits

Cover Photo: © David Woodfall/Getty Images
Anthony Annandono, 14, 54
© AP/Wide World Photos, 65
© Bettmann/CORBIS, 11, 19, 62
© Bloomberg News/Landov, 66
© Mark Boulton/Photo Researchers, Inc., 17
© Andrew Brown; Ecoscene/CORBIS, 25
© Clouds Hill Imaging Ltd./CORBIS, 31
© Robert Estall/CORBIS, 37
© Simon Fraser/Photo Researchers, Inc., 39
© Hulton/Archive by Getty Images, 49
© D. Lovegrove/Photo Researchers, Inc., 52
© Steve Maslowski/U.S. Fish and Wildlife Service, 46
© Will & Deni McIntyre/Photo Researchers, Inc., 42
© Sally A. Morgan; Ecoscene/CORBIS, 26
© T. Nilson/Jvz/Photo Researchers, Inc., 7
© Charles Philip/CORBIS, 58
PhotoDisc, 9, 29
© Pete Saloutos/CORBIS, 56
© Ted Spiegel/CORBIS, 44, 69
© Joseph Sohm; ChromoSohm Inc./CORBIS, 70
© Paul A. Souders/CORBIS, 21
© Roger Tidman/CORBIS, 34

About the Author

Rebecca K. O'Connor works at a desert zoo in Southern California, where everyone understands the importance of rain. With a zoo background, she has an interest in all living creatures and the ecosystems in which they live. She has also written *Frogs and Toads* and *Owls* from the Endangered Animals and Habitat Series for Lucent Books.

DATE DUE

HIGHSMITH 45230